Evening Pr

Its remarkable properties and its
range of conditions.

By the same author
MULTIPLE SCLEROSIS — A Self-help Guide to its
Management
THE Z FACTOR — How Zinc is Vital to Your Health
(with Dr Michel Odent)

EVENING PRIMROSE OIL

Second Edition

by
JUDY GRAHAM

THORSONS PUBLISHING GROUP

First published March 1984
Second Edition 1988

British Library Cataloguing in Publication Data

Graham, Judy
Evening Primrose Oil.
1. Evening Primrose Oil — Therapeutic use
I. Title
615'.323'672   RS165.B/

ISBN 0-7225-0743-7

Published by Thorsons Publishing Group Ltd.,
Wellingborough, Northamptonshire, NN8 2RQ, England

Printed in Great Britain

3   5   7   9   10   8   6   4   2

This book is dedicated to the doctors who are working to solve the many diseases of civilization.

'Only connect'
E.M. Forster.

The Evening Primrose

by John Clare

When once the sun shrinks in the west,
And dew-drops pearl the evening's breast;
Almost as pale as moonbeams are,
Or its companiable star,
The evening primrose opens anew
Its delicate blossoms to the dew
And, hermit-like, shunning the light,
Wastes its fair bloom upon the night;
Who, blind-fold to its fond caresses,
Knows not the beauty he possesses.

Thus it blooms on while night is by;
When day looks out with open eye,
'bashed at the gaze it cannot shun,
It faints and withers and is gone.

CONTENTS

ACKNOWLEDGEMENTS

My thanks are due to the many doctors who have done the research work on evening primrose oil. I have quoted freely from their published papers in order to keep as closely as possible to their own accounts of their findings. Their names are too numerous to mention here. They appear in the references at the back of the book.

Special thanks are due to Valerie Grundon, Howard Thomas, Peter Lapinskas, David Dow, Leslie Smith and John Williams who all gave me of their valuable time to help me in the research on evening primrose oil.

Much of the research work written about in this book has been funded by Efamol Ltd. My particular thanks are due to Dr David F. Horrobin, Chief Executive of Efamol, for letting me have all the research papers without which this book could not have been written, and to Christina Toplack, Research Librarian at the Efamol Research Institute, who gave me such painstaking help with all the references.

Note to readers

Before following the self-help advice given in this book, readers are urged to give careful consideration to the nature of their particular health problem and to consult a competent physician if in any doubt. This book should not be regarded as a substitute for professional medical treatment and whilst every care is taken to ensure the accuracy of the content, the author and the publishers cannot accept legal responsibility for any problem arising out of the experimentation with the methods described.

PREFACE

It might seem very odd that someone who is neither a scientist nor a botanist should be writing a book about evening primrose oil. I even find it quite surprising myself that I have become so involved in this bizarre little plant and its fantastic possibilities.

My own involvement with evening primrose oil goes back to the early 1970s, when only a very few people had ever heard of it. It was then that I began taking it for multiple sclerosis and have carried on ever since.

In my book *Multiple Sclerosis — A Self-help Guide to Its Management* (Thorsons, revised edition 1987) I devoted a whole chapter to evening primrose oil, and this book came to be written as a development of that chapter.

I also became fascinated by all the marvellous possibilities of evening primrose oil, and followed the various scientific studies with great interest. Over the years of patient waiting, many of the early hypotheses about evening primrose oil have been proved correct in scientific trials.

Even though I take the oil myself, this book has been written with the objectivity required of any journalist. I am simply reporting the facts and letting them speak for themselves.

INTRODUCTION

The story of the evening primrose could turn out to be one of the most fantastic in plant and medical history. Like foxglove (digitalis), cinchona bark (quinine) and rauwolfia (reserpine) before it, the evening primrose promises to take its place in the hall of fame of plants with important medicinal properties. But unlike these other natural products which are, on the whole, useful for only one condition, the oil of the evening primrose has properties which make it useful for a very wide range of illnesses.

A sceptic might think he'd strayed into a Victorian fairground where some charlatan was selling strange mixtures promising to cure everything from gout to hiccups with one gulp. And, on the face of it, the claims made for evening primrose oil do seem quite fantastic. After all, how can the same unassuming little plant be used in the treatment of benign breast disease *and* brittle nails *and* faulty blood vessels, and a score of other ailments?

The answer is that, although all these conditions seem very different, evening primrose oil has something in it which is needed in each of them. Evening primrose oil is a very rich source of essential fatty acids, which work a bit like vitamins in the body. And evening primrose oil has a very rare ingredient in it called gammalinolenic acid (GLA) which sets it apart from most other oils.

GLA is very special indeed. It does two quite distinct things which make it so useful for so many different health conditions. First, it helps build healthy cell membranes in every single cell of the body. And secondly, GLA converts inside the body to a physiologically-active substance called Prostaglandin E1 (PGE1). GLA and PGE1 together are the real heroes of the piece.

Recently, this rare GLA has also been found in oats, barley, borage, blackcurrant oil, and in a specially fermented mould. It is also available in small amounts in cow's milk. The fresh water alga, spirulina, also contains GLA. So does anchusa, rose bay willow herb and comfrey, but in these plants the GLA is not readily available for human consumption. A rich source of GLA is human breast milk, which is one reason why breastfeeding is so important for babies.

Many products containing GLA are now on the market, some of them combined with evening primrose oil. Some of these products contain high amounts of GLA, indeed higher than the GLA percentage in some evening primrose oil capsules (9% in *Efamol*). However, the bigger amount of GLA in some of these other products does not necessarily mean they work better. In some conditions they may be far less effective than evening primrose oil. This may have something to do with the detailed chemical composition of the different oils.

None of these other products containing GLA has been tested in a scientific way on humans in the way that evening primrose oil has. So the results written about in this book refer only to evening primrose oil, even though some other products with GLA might work in a similar way.

Evening primrose oil is widely available in health food shops and chemists as a dietary supplement of essential fatty acids. It has been the subject of more than 100 scientific trials in universities and hospitals all over the western world, and many more are still in progress. It is very unusual indeed for a health food product to be subjected to such stringent scientific study. In many scientific papers about evening primrose oil it is referred to as a 'drug'. Certainly, evening primrose oil has pharmacological properties. This book tells you what they are and how they work in many different conditions.

1 WHAT IS THE EVENING PRIMROSE?

A few years ago, no one had ever heard of the evening primrose. But in the last few years it has caught the public's imagination and become one of the most popular products in health food stores.

The profile of the evening primrose has changed dramatically in the last decade. Before then, it was a humble weed straggling along waysides. Now, farmers faced with the need to diversify are planting acres of this latest cash crop.

The evening primrose has come a very long way from its wild days along roadsides, railway sidings, waste sites and sand dunes.

Strictly speaking, the evening primrose is not a primrose at all. It is related to the rose bay willow herb family, and to the popular garden flowers clarkia and godetia.

It acquired its name because its bright yellow flowers look like the colour of real primroses, and because its flowers open in the evening. It has the curious habit of blooming between 6 and 7 o'clock in the evening, when eight or ten of the largest fragrant flowers can burst open every minute. The flower usually lasts for the whole of the next day, particularly in dull weather, but in bright sunlight the flowers fade quite quickly. In England, the plant flowers from the end of June to mid August.

Experts who classify plants (taxonomists) will tell you that the evening primrose belongs to the order *Myrtiflorae*, family *Onagraceae*, genus *Oenotherae*. The generic name comes from the Greek *oinos* (wine) and *thera* (hunt). According to herbals, this described a plant — probably a willow herb — which gave one a relish for wine if the roots were eaten. Another interpretation is that the plant dispelled the ill effects of wine, and this fits in better with modern research (see Chapter 13).

Herbals describe the evening primrose as being astringent and sedative, and the oil helpful in treating gastro-intestinal disorders, asthma, whooping cough, female complaints, and wound healing.

History

Research has revealed that the species originated in Mexico and Central America some 70,000 years ago. Four times it colonized North America, each time being almost wiped out by glaciers during successive ice ages. Each new wave of evening primrose cross-pollinated with survivors and so continued the line.

American Indians are supposed to have used the evening primrose for hundreds of years. According to folklore, a tribe called Flambeau Ojibwe were the first to realize the medicinal properties of the evening primrose plant. They used to soak the whole plant in warm water to make a poultice to heal bruises, they used the plant for skin problems and asthma, and brewed a cough mixture from the roots.

From America, the evening primrose spread all over the world. Botanists first brought the plant from Virginia to Europe in 1614 as a botanical curiosity.

Most of the strains, however, came to Britain during the next century as stowaways in cargo ships carrying cotton. As cotton is light, soil was used as ballast. The ballast was dumped on reaching port, and with it stray seeds of evening primrose. Even today there are areas around the major ports, such as Liverpool, where evening primrose plants — descendants of the cotton ballast — grow in profusion.

In Europe, the evening primrose became known as 'King's Cure All' by those who knew its almost magical medicinal properties. For centuries, however, the evening primrose was left to straggle along without anyone but a few specialist herbalists taking much notice. It wasn't until this century that scientists began to look at the plant for its industrial potential in such things as paint.

In 1917 a German scientist called Unger examined the plant, and found that the seeds contained 15% oil, which was extractable with light petroleum.[1] In 1919 the Archives of Pharmacology published a paper by Heiduschka and Luft who were the first to do a detailed analysis of the oil. They extracted 14% oil with ether, and apart from the normal oleic and linoleic acids, found a new fatty acid, which they named gammalinolenic acid (y-linolenic acid).[2] In 1927, three German scientists repeated the Heiduschka and Luft test, and came up with a more detailed analysis of the chemical structure of this gammalinolenic acid (GLA).[3]

Twenty-two years later Dr J.P. Riley, a British biochemist in the Department of Industrial Chemistry at Liverpool University, came across the German papers on evening primrose oil and decided to analyse the oil for himself, but this time using modern techniques. So Dr Riley set off for the sandhills near Southport in Merseyside and picked a bunch or two of evening primrose plants. He dried the plants, separated the seeds, and extracted the oil. To his great satisfaction, he found for himself the unique gammalinolenic acid.[4]

It wasn't until the 1960s, however, that British scientists began investigating the oil for its possible health uses. The first experiment was on rats. The aim of this experiment was to compare the biological activity of the commonly-found linoleic acid with the rare gammalinolenic acid.[5]

The rats were put on a diet lacking in essential fatty acids, and after a few weeks they developed loss of hair and skin problems. They were then divided into two groups. One group was fed linoleic acid and the other group was fed gammalinolenic acid. The results of this first experiment were remarkable. The rats in the GLA group recovered more rapidly than the other group, and there was evidence that the GLA was far more efficiently taken up by the cells of all the important tissues and organs of the body.

The success of this experiment encouraged biochemists and other scientists to do more research work on this unusual seed oil. The next test was to investigate the effect of the GLA in regulating cholesterol production in a group of rabbits being fed a diet high in animal fats and to compare the results with a control group being fed a normal diet. The results showed that the GLA in evening primrose oil could control blood cholesterol levels. Further experiments followed, opening up yet more possibilities, particularly in heart disease research.

Much of this research during the 1960s was the brainchild of biochemist John Williams. If it had not been for this one man, evening primrose oil might well have been lost in the archives of obscure journals forever. At the time, John Williams was working for a small firm. But a large pharmaceutical company took over this small firm, and they had a policy of not getting involved in natural products. They decided to drop the seed oil project. John Williams saved the project — his 'baby' — by taking an early retirement, and taking the seed oil project elsewhere. During the year after he left the company, John Williams could not get the evening primrose oil project off his mind. Fortunately, with the help of a Cheshire businessman he was able to take over the project and buy all the seed stock, documents and patents and start up his own company, Bio-Oil Research Ltd. They began to manufacture capsules of evening primrose oil under the brand name *Naudicelle*, and carried on with their scientific studies into the benefits of the oil.

During the 1970s, many other scientists took a keen interest in the possibilities of evening primrose oil. Dr Ahmed Hassam and Michael Crawford (now Professor Crawford) did various experiments at the Nuffield Laboratories of Comparative Medicine in London which showed the GLA was ten times more biologically effective than linoleic acid.[6,7]

The first disease for which evening primrose oil was used was multiple sclerosis (see Chapter 12). This followed a paper by some eminent neurologists which said that sunflower seed oil reduced the severity and frequency of relapses in the disease.[8] Dr Hassam's evidence suggested that if sunflower seed oil helped a little, then evening primrose oil, being that much more active, could help even more.

A leading specialist in MS, Professor E.J. Field, who at that time was at Newcastle University, tested the oil in 1974 and did find an increased activity compared to ordinary sunflower seed oil. On Professor Field's advice, many people with MS have been taking evening primrose oil capsules ever since.

A neurophysiologist who at one time worked in the same university as Professor Field was responsible for catapulting evening primrose oil into the wider fields of international medical research. Dr David Horrobin, at that time Reader in Physiology at Newcastle University Medical School, first became interested in evening primrose oil in the mid 1970s, as a result of his work on schizophrenia and prostaglandins.

Dr Horrobin, with Dr Abdulla of Guy's Hospital, London, discovered that schizophrenics have extremely low levels of Prostaglandin E1 (PGE1) which is manufactured in the body from gammalinolenic acid.[9] This finding also set Dr Horrobin thinking about the role of PGE1 in many other diseases.

This enterprising young doctor was so convinced of the far-reaching possibilities of gammalinolenic acid that he persuaded a director of Agricultural Holdings Ltd (which included the seed merchants Hurst Gunson Cooper Taber Ltd who had supplied John Williams with his seeds) to set up another company to research, develop and market evening primrose oil. This new company was Efamol Ltd.

Bio-Oil Research Ltd and Efamol Ltd were two of the first companies involved in evening primrose oil. Several other companies now make evening primrose oil. However, it was Dr David Horrobin, now President and Research Director of Efamol Ltd, who has been behind most of the medical trials concerning evening primrose oil. Efamol Ltd has been at the forefront of the research involved in the therapeutic possibilities of evening primrose oil.

Today, research on evening primrose oil is still going on, with several studies in progress and more in the pipeline.

2 FROM WILD FLOWER TO CROP

In its natural state, the evening primrose is a wild flower. But it is only in the last decade or so that agronomists have turned this wild, untamed flower into a crop.

Most crops have been in existence for centuries. Farmers have had plenty of time to learn about their cultivation. The evening primrose, however, is a brand new crop and has presented an enormous challenge to agronomists who are trying to master its cultivation in a fraction of the time devoted to other crops.

Grown from seed, the evening primrose plant produces a rosette of leaves close to the ground in the first year. The following season the plant shoots up to produce a main stem that can be 5 or 6ft tall (almost 2m) bearing the attractive yellow flowers, and then the seed pods in late summer and early autumn.

A unique botanical specimen

Botanists first became interested in the evening primrose plant at the beginning of the century when it was thought it was breaking the laws of inheritance discovered a few years earlier by Mendel.

In ordinary species, if two plants are cross-pollinated, the first generation of offspring are identical to one another. But in the second generation they form a mixture of two types, intermediate between the two parents. In further generations, new variants continue to be produced.

The evening primrose, however, does something very different. If two plants are crossed, the first generation usually turns up as a mixture of two types which don't resemble either of the two parents or each other. In the next generation, the two

groups do not split up into a mixture. Instead, the plants breed
true.[1] This is because the plants have an unusual chromosome
which can repeat itself for many generations without any
variation.

From a distance, a cluster of plants might look the same. But
at a closer look one sees there are some subtle variations between
one type and another. The size of the flowers can vary from ½in
to 5in (1 to 13cm), and the pigment may differ in each plant.
Although most evening primroses are yellow, some have a
mauve hue, and there are many variations in colour in each part
of the plant. There are huge differences in plant size, too. Some
grow no higher than dandelions, while others can shoot up to
more than 8ft (2.4m) high.

Some evening primrose plants need bees or moths for
pollination; others are self-pollinating. The variations provide
great challenges for plant breeders aiming to breed the evening
primrose for its oil yield.

Breeding evening primrose plants for seed oil

Plant breeders are creating new hybrids of evening primrose
plants by means of carefully controlled hand cross-pollination of
the best available plants.

Each parent plant selected for hybridization carries some of
the target characteristics of the ideal variety. This process takes
years of observing closely to see how each new hybrid performs
and carefully measuring the oil yield after the cross-pollination.

When a new hybrid is produced that looks promising, it must
be reproduced and multiplied for several generations before
anyone can be sure that a stable new variety which gives
consistent oil yields of a high quality has been produced.

With plants grown for their seeds, the new hybrid had to be a
crop which could germinate more quickly and evenly than the
wild types. In its wild state, germination is unreliable. In fact,
the evening primrose plant has many unsuitable characteristics
in its wild state. Some of the wild evening primroses can shoot
up, or 'bolt' up, too early and be cut down by the winter frosts or
else fail to grow quickly enough in the spring and produce no
seed. So another requirement with a hybrid was to develop a
plant in which the tall stems bearing the flowers, and later the
seed pods, should start to grow neither too early, nor too late.

The aim of the plant breeders has been to produce a hybrid

with the most consistent yield of GLA. At the moment the GLA content in the oil of the hybrids is around 9%. This is the oil composition on which almost all the human research has been performed.

The highest yield of GLA comes when the seeds are mature. Another weakness of the wild plant is that the flowers of the evening primrose bloom in succession over a period of two or three months. Seed pods form each day, which means that a single plant could well be carrying some mature pods, unripe pods, freshly opened flowers, and even immature flower buds. So plant breeders who have been working on the evening primrose hybrids have been aiming for plants carrying the maximum number of mature pods at a given time for the best possible seed yields and the highest GLA content in the oil.

In its wild state, the evening primrose yields an oil which varies greatly in quality and the amount of GLA is hopelessly inconsistent, some types giving far too little to be effective. So it has only been by a rigorous plant breeding programme that a few new varieties have been developed which have all the characteristics needed for a seed oil crop and which give a consistently reliable high quality GLA yield of around 9%.

Completely new varieties of evening primrose, named Constable, Cossack and Commodore have been specifically bred by Efamol and that company has been granted Plant Breeders' Rights — the equivalent of a plant patent — in the UK and the USA.

Now, new hybrids of evening primrose are being grown in various parts of Britain, in Spain, Hungary, the USA, and other parts of the world. When farmers want to diversify, and move into more unusual crops than wheat or corn, they are choosing the evening primrose.

A bonus for farmers is that the rate of multiplication of the evening primrose is very high — one kilo of seed multiplies into 1000 kilos for the farmer. But then it takes 5000 of the tiny seeds to make just one capsule of evening primrose oil!

Quality control

A uniform seed is used, and farmers are advised on how to get the best results from the evening primrose crop. Batches of seed are routinely analysed to make sure no contaminants have got in, and that the GLA content conforms to the specification.

The oil is routinely analysed and checked, and similar stringent checking goes on during the extraction and encapsulation processes, to make sure of a pure, high quality yield with exactly the same amount of oil in each capsule.

Evening primrose oil is maintained in the same state as it occurs naturally in the seeds: a clear, pale yellow, natural oil with no additives, colouring or other processing. It is sold in clear gelatin capsules, and the only thing that is added in some brands is Vitamin E, which prevents oxidation of the oil.

3 ESSENTIAL FATTY ACIDS

Evening primrose oil is a very rich source of essential fatty acids, which are as important to the body as vitamins, minerals or proteins.

Essential fatty acids are vitamin-like lipids. They are called 'essential' because the body must have them and can't make them by itself, so you have to eat the foods containing them.[1,2] Some people have called this type of essential fatty acid 'Vitamin F' as there are some similarities with vitamins.

Essential fatty acids have got nothing to do with fattiness in the sense of pork dripping or cream buns. In fact, you will find essential fatty acids in foods that don't look remotely 'fatty' at all, like sunflower seeds, green vegetables, and shellfish.

Fatty acids are an essential part of nutrition and they perform all kinds of vital functions within the body:

- They give energy
- They help maintain body temperature
- They insulate the nerves
- They cushion and protect tissues
- They are part of the structure of every cell in your body and are vital for metabolism
- They are also precursors of the all-important short-lived regulating molecules, the prostaglandins (see Chapter 4)[3]

The Food and Agricultural Organization suggests that a minimum of 3% of total calories should be from essential fatty acids in adults, and 5% in children and pregnant and breastfeeding women.[4] Roughly 60% of the brain is made up of lipids, of which an important part is essential fatty acids. They are vital for the proper growth and development of the brain and the central nervous system.

Essential fatty acids (EFAs) have two major roles. First, they are constituents of all cell membranes and in all tissues in the body. And second, they give rise to a group of highly reactive molecules, the prostaglandins and leukotrienes (see next chapter).

How fluid and flexible cell membranes are depends on how many EFAs they have. This influences the behaviour of every cell in the body including nerve cells and blood cells, the lymphocytes and neutrophil leukocytes, which need to be strong and healthy as their job is to rid the body of foreign invaders. The activity of these lymphocytes depends on the state of the cell membranes. They will behave differently according to whether the cell membrane is fluid and flexible (i.e. plenty of EFAs) or rigid (not enough EFAs).

Evening primrose oil has beneficial effects in all areas of the body both because of its actions on cell membranes, and because it is a precursor of a particularly helpful prostaglandin. These beneficial effects can best be seen by looking at what happens in EFA deficiency states.

Essential fatty acid deficiency

Growing animals who are totally deprived of all essential fatty acids show a wide range of unpleasant symptoms.[1,2,5] These are the main ones:

- The hair falls out and eczema-like skin lesions develop. The sebaceous glands hypertrophy.
- Wounds fail to heal normally, apparently due to a failure of connective tissue formation.
- Normal growth fails to occur.
- All body membranes become exceptionally permeable. In particular the skin loses its ability to prevent passage of water. Large amounts of water are lost across the skin, the animal is thirsty, but the urine is concentrated.
- Reproduction fails, especially in males. Females may become pregnant but frequently miscarry and rarely carry a litter to term.
- The kidneys hypertrophy and are prone to haemorrhage, and renal failure develops.
- The liver undergoes fatty degeneration.
- The tear and saliva glands atrophy and so does the pancreas.

• The immune system is defective and there is great susceptibility to infections.

The earliest results of such experiments were obtained as long ago as 1929.[6] Of course, there have been no such deliberate experiments on humans. However, when baby milk formulas were first being developed, they left out EFAs and the babies' skin became dry, scaly and eczema-like.[7,8] A similar thing happened when hospital patients were first fed totally by intravenous drip solutions. The drip solutions left out EFAs, and the patients developed skin rashes resembling eczema or psoriasis.[9] In both cases, the symptoms cleared up once EFAs were given.

Although there have been no long-term studies on EFA deficiency states in humans, one way to test for such deficiency is to see what happens to symptoms when supplemental amounts of EFAs are given.

The families of essential fatty acids

There are two families of essential fatty acids. One is the linoleic acid family, and the other is the alpha-linolenic acid family.

Linoleic acid is found in dairy products, organ meats such as liver, human milk, and notably in vegetable seed oils such as sunflower, safflower, and corn.

Alpha-linolenic acid is found in foods such as green vegetables, soya, and linseeds. Fish, fish oils and shellfish contain the metabolites of alpha-linolenic acid, eicosapentaenoic acid and docosahexaenoic acid. Both these families of essential fatty acids are vital for good health.

Evening primrose oil contains EFAs belonging to the linoleic acid family. There is no shortage of linoleic acid in foodstuffs. In modern diets, milk products and vegetable seed oils together with foods derived from them are probably the most important sources. In Europe, the average adult intake is around 7-15g a day. In North America, it is around 10-25g a day.

So, if there is no shortage of linoleic acid in the diet, why should anyone need to take a supplement like evening primrose oil?

THE PROBLEM IS THAT THE ESSENTIAL FATTY ACIDS IN THE FOOD YOU ARE EATING MAY NOT BE GETTING THROUGH TO THE PLACES THEY'RE NEEDED MOST.

Figure 1. Essential fatty acid metabolism.

Linoleic acid is the parent of one family, and alpha-linolenic acid is the parent of the other family.

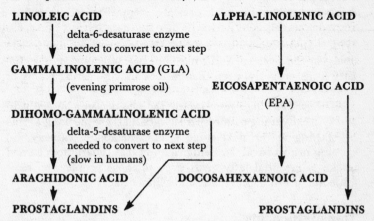

Blocking agents

Linoleic acid on its own has relatively little biological activity. For most of its uses it has to be converted in the body to other things which are biologically active.

In order for linoleic acid to be converted to the next stage, gammalinolenic acid (GLA), a particular enzyme is needed which is called delta-6-desaturase. However, there are many things which can prevent linoleic acid from being converted to GLA. This can happen either because the key enzyme is defective, or the enzyme's action is inhibited, or because the linoleic acid is in a form which cannot be converted at all.

These are the most common agents which block GLA formation:

• Foods rich in saturated fat
• Foods rich in cholesterol
• Foods rich in trans fatty acids

- Alcohol, in moderate to large amounts
- Diabetes (too little insulin)
- Too much sugar
- Stress hormones (the catecholamines such as adrenaline)
- Ageing
- Zinc deficiency
- Viral infections
- Radiation
- Cancer
- Atopic conditions

With so many blocking agents[6,7,8] it's not surprising that people living in the western world may not be metabolizing essential fatty acids properly, even though they may be eating plenty of them.

All these blocking agents inhibit the step between linoleic acid and GLA. Evening primrose oil avoids all these blocking agents by starting at the next stage in the metabolic pathway. So evening primrose oil, by being rich in GLA, guarantees a good intake of essential fatty acids even in those people who have impediments to the metabolism of linoleic acid.

Trans fatty acids

The only type of linoleic acid which can convert into biologically useful things is what is known as cis-linoleic acid. This is when the oil (e.g corn oil, sunflower seed oil, etc.) is in its natural, unadulterated state.

Only cis-linoleic acid has any real value as an essential fatty acid. Evening primrose oil contains over 70% cis-linoleic acid, as well as about 9% cis-gammalinolenic acid, and none of this is lost when the oil is put into capsules.

Once linoleic acid and alpha-linolenic acid get processed during food manufacturing they may be turned into biologically different forms which are real non-starters as far as the metabolic pathway is concerned. This is because what were originally biologically active essential fatty acids have been turned into biologically inactive trans fatty acids.

Trans fatty acids are lurking in everyday foodstuffs. On the face of it, they look perfectly pleasant and often delicious — ordinary bottles of cooking oil on supermarket shelves, cartons of margarine, luscious looking pastries, lovely sweets, tasty french fries, and all those foods which people generally love to eat. But beware! Trans fatty acids behave as if they were saturated fats. And far from being essential fatty acids, they actually produce EFA deficiency states and compete with genuine EFAs for your body's time and attention. They elbow the real EFAs out of the action.[10,11,12,13,14,15,19]

The trans fatty acids which you eat make their way into tissues like the brain, heart and lungs, and some scientists are sure that they change the properties of these tissues — probably for the worse.[16,17,18] Trans fatty acids are probably safe in normal, healthy humans but they may not be in ill people who suffer from the conditions listed in this book, or in people with a defective D6D enzyme.

Food processing techniques have particularly hit the alpha-linolenic acid family. Unfortunately, because of its instability, food manufacturers usually try and stabilize it by hydrogenating it, which gives it a longer shelf life. But this leads to loss of its EFA activity.

It has been argued that the fall in intake of biologically active EFAs of the alpha-linolenic acid family has been one of the major changes in western nutrition in the last 50 years. Trans fatty acids also depress the formation of prostaglandins, which suggests that they have an effect not just at the first step in the metabolic pathway, but at later stages too.

Here is a table from the United States which shows the percentages of trans fatty acids in foodstuffs.[20]

- Bakery products — up to 38.5%
- Sweets — up to 38.6%
- French fries — up to 37.4%
- Hard margarines — up to 36%
- Soft margarines — up to 21.3%
- Diet margarines — up to 17.9%
- Vegetable oils — up to 13.7%
- Vegetable oil cooking fats — up to 37.3%

The cockeyed thing is that when government and other bodies involved in the nation's nutrition examine the amounts of lipids (fats) people are eating, they often lump together both trans and cis fatty acids, perhaps not being fully aware that trans fatty acids may as well not be counted for all the nutritional good they do. This means that the overall intake of real essential fatty acids is lower than we have been led to believe.

It's interesting to ponder the fact that it's only since the 1920s that significant amounts of trans fatty acids began to be added to the diet (though they have always existed in small amounts in dairy produce). People who are interested in the geographical distribution and the increase in diseases of western civilization might find this worthy of more research. Trans fatty acids are by no means the only blocking agents, but they are the ones which cause innocent eaters to fall into the trap of thinking they're eating the right things when in fact they're eating the wrong things.

Polyunsaturated fatty acids and essential fatty acids

All essential fatty acids are polyunsaturated fatty acids (PUFAs). Whether fat is saturated, unsaturated, or polyunsaturated has to do with its biochemical composition.

Fat is made from smaller components called fatty acids. In biochemical terms, fatty acids are chain-like substances, some with short chains, and some with long chains. The chains are of carbon atoms with hydrogen and oxygen atoms attached. The degree of saturation depends on the extent to which they can absorb more hydrogen.

Although all essential fatty acids are PUFAs, the reverse does not apply. Most PUFAs are in fact *not* EFAs.

People who are in the business of promoting good health often say that you should increase your intake of PUFAs. But this is not entirely sound advice, and can be downright bad advice. The beneficial effects of PUFAs appear to be due entirely to their EFA content. If the linoleic acid is in the cis-form, the PUFAs are also EFAs. But if the oils are in the trans-form, they should be regarded as saturated fats as far as their health risks are concerned.

Doctors who recommend to their patients to switch from butter to margarine may not realize that by doing so the patients may actually be increasing their ratio of potentially harmful fats. Most margarines and vegetable cooking fats contain 5 to 50% of trans PUFAs. So read labels very carefully. Only buy margarines which state that the cis bonds are intact. Avoid anything hydrogenated.

Vitamin and mineral co-factors

The metabolism of evening primrose oil works best if it is taken together with some specific vitamin and mineral 'co-factors'.

The particular vitamins and minerals which help evening primrose oil along its metabolic journey are:

- Vitamin C
- Vitamin B6
- Nicotinamide (also known as Vitamin B3 or niacin)
- Biotin
- Zinc
- Magnesium
- Calcium

Also, to prevent oxidation of the evening primrose oil, it MUST be taken with Vitamin E. Some brands of evening primrose oil including *Efamol* and *Naudicelle* already contain Vitamin E.

Evening primrose oil and fish oils

In recent years, scientists have begun to understand the importance of the two families of essential fatty acids working together — the linoleic acid family combined with the alpha-linolenic acid family. The best combinations seem to be the metabolites, or derivatives, of the parent EFAs, rather than the parents themselves. The best combination seems to be GLA (as in evening primrose oil) and eicosapentaenoic acid (EPA) as in fish oils.

Evening primrose oil is now available combined with fish oils. For example, *Efamol Marine* has a mixture of 80% evening primrose oil and 20% concentrated fish oil in 500mg capsules. The ratio of the linoleic acid family to the alpha-linolenic family should be in the range of between 3 and 6 to 1. Put simply, this means that for every 3, 4, 5, or 6 capsules of evening primrose oil, you should take one containing alpha-linolenic acid, eg fish oil. (The capsules which contain both evening primrose oil and fish oil are of course already in the right ratio.)

It is probable that for some conditions a combination of GLA and EPA would give better results than GLA on its own, and so the combined oils have been tried recently in some medical trials (see Chapter 10). On the other hand, for some conditions GLA alone may give the best results. Many trials are now going on to sort out which is which.

Figure 2. The metabolic conversion of cis-linoleic acid

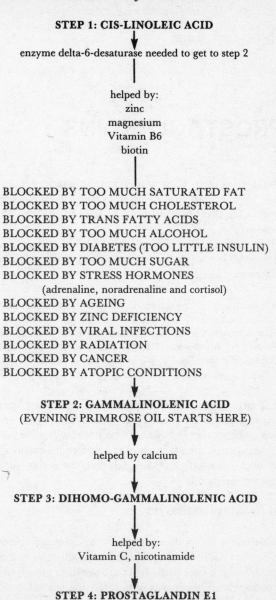

STEP 1: CIS-LINOLEIC ACID

enzyme delta-6-desaturase needed to get to step 2

helped by:
zinc
magnesium
Vitamin B6
biotin

BLOCKED BY TOO MUCH SATURATED FAT
BLOCKED BY TOO MUCH CHOLESTEROL
BLOCKED BY TRANS FATTY ACIDS
BLOCKED BY TOO MUCH ALCOHOL
BLOCKED BY DIABETES (TOO LITTLE INSULIN)
BLOCKED BY TOO MUCH SUGAR
BLOCKED BY STRESS HORMONES
    (adrenaline, noradrenaline and cortisol)
BLOCKED BY AGEING
BLOCKED BY ZINC DEFICIENCY
BLOCKED BY VIRAL INFECTIONS
BLOCKED BY RADIATION
BLOCKED BY CANCER
BLOCKED BY ATOPIC CONDITIONS

STEP 2: GAMMALINOLENIC ACID
(EVENING PRIMROSE OIL STARTS HERE)

helped by calcium

STEP 3: DIHOMO-GAMMALINOLENIC ACID

helped by:
Vitamin C, nicotinamide

STEP 4: PROSTAGLANDIN E1

4 PROSTAGLANDINS

Prostaglandins are very important substances in the body because they act as vital regulators almost everywhere. Prostaglandins are fast becoming one of the biggest growth areas in biochemistry.

This is what Dr David Horrobin, who has done much of the frontier-breaking work, says about the revolutionary possibilities of prostaglandins:

> I believe that in the next decade we in the prostaglandin field have the opportunity to bring about a revolution in both biology and practical medicine which will have a greater impact on people's lives and on fundamental biological concepts than any previous bio-medical revolution. Few can doubt that the prostaglandins will have an impact approaching that of the antibiotics 30 years ago.[1]

Evening primrose oil is a 'precursor' of one type of prostaglandin, prostaglandin E1 (PGE1). It is the PGE1, together with the fluidizing effect on membranes, which is responsible for so many of the therapeutic effects in so many different conditions. PGE1 is synthesized easily from GLA, which is the active ingredient of evening primrose oil.

What prostaglandins do

Prostaglandins were discovered by a Swedish scientist, von Euler, in the 1930s. He first found these molecules in the seminal fluid and thought they came from the prostate gland, so he called them prostaglandins. It was only later that scientists discovered that these molecules were all over the body.

Prostaglandins act as vital cell regulators. They control every cell and every organ in your body on a second-by-second basis. The nearest thing to them is hormones, which also have important messenger roles. But prostaglandins aren't like hormones, which zip around all over the place. Prostaglandins are much more local than that; they're a bit like neighbourhood hormones regulating everything but staying on their own patch.

Each prostaglandin has a very specific effect in each tissue. The main actions of prostaglandins seem to be as local messengers which regulate the activity of the tissues in which they are produced.[2] They also regulate the activity of certain key enzymes. This vital job as a cell regulator helps explain why one hero of this book, PGE1, keeps on popping up in every chapter. And because the prostaglandins have such a key role as regulators of every cell, it helps to explain why this important molecule could possibly be of such vital importance in so many apparently different conditions.

Since von Euler's discovery in the 1930s, prostaglandins have been found particularly in blood vessel walls,[3] macrophages,[4] platelets,[5] duodenal secretions,[6] nerves,[7] and every organ.

Prostaglandins have an extremely short life span; most are removed from the blood during a single passage through the lungs. They are so short-lived in part because they are naturally unstable, and in part because they have highly efficient mechanisms which break them down. The very short life span of prostaglandins make them difficult to administer as drugs because they have to be given intravenously.

Types of prostaglandin

There are three series of prostaglandins: PG1, PG2, and PG3. Each of these has a different chemical structure. And within each series there are many different types, classified by letters A, B, D, E, F, etc. In all, there are at least 50 prostaglandins, with new ones being discovered each year.

The three series of prostaglandins are each derived from a different fatty acid. Series 1 and 2 both come from the linoleic acid family. Series 3 is derived from eicosapentaenoic acid, a member of the alpha-linolenic acid family and most commonly found in oily sea foods.

Each prostaglandin has a different role to play. Health problems can arise when the different series of prostaglandins are out of balance with each other. The balance between the 1 and 2 series PGs can be influenced by diet. In inflammatory conditions, the end-products of arachidonic acid metabolism — prostaglandins series 2, cyclo-oxygenase and thromboxane A2 — are being produced in too great a quantity, whereas PGE1 is not being produced in great enough quantities.

Two of the most widely-used drugs — steroids and Non-Steroidal Anti-Inflammatory Drugs (NSAIDs) work by inhibiting the biosynthesis of prostaglandins. In orthodox medicine, many of the conditions listed in this book are treated with these drugs. By suppressing the production of prostaglandins, the drugs dampen down inflammation.

The trouble with this drug approach is that *all* the prostaglandins are knocked out — including the good ones. Evening primrose oil works in a completely different way from these powerful drugs. Instead of stopping the manufacture of prostaglandins, evening primrose oil goes on to make the anti-inflammatory prostaglandin E1, and so it manipulates the prostaglandins but in a completely natural way.

Prostaglandin E1 (PGE1)

Of all the prostaglandins researched so far, PGE1 seems to have the most highly desirable qualities. Evening primrose oil is easily converted in the body to prostaglandin E1.

These are just some of the beneficial things that PGE1 does in the body:

- It inhibits dilation of blood vessels.
- It lowers arterial pressure.
- It inhibits thrombosis.
- It inhibits cholesterol synthesis.
- It inhibits inflammation and experimental arthritis.
- It inhibits abnormal cell proliferation[3].
- It inhibits platelet aggregation[1].
- It regulates production of saliva and tears.
- It elevates cyclic AMP (adenosine monophosphate).

Evening primrose oil, because it converts in the body to PGE1, has been tested in all the conditions listed in this book because of what was known about the wonder-workings of PGE1.

Figure 3. How prostaglandin E1 works in the body.

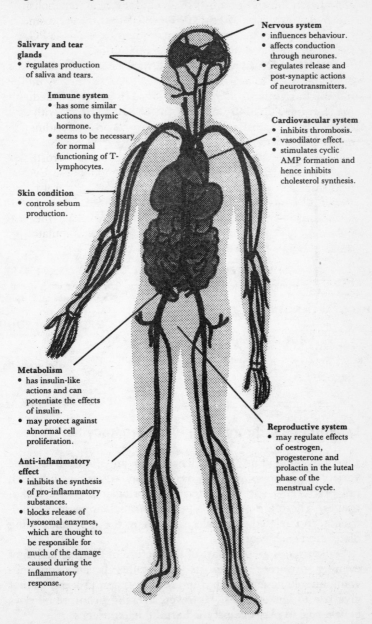

Nervous system
- influences behaviour.
- affects conduction through neurones.
- regulates release and post-synaptic actions of neurotransmitters.

Salivary and tear glands
- regulates production of saliva and tears.

Immune system
- has some similar actions to thymic hormone.
- seems to be necessary for normal functioning of T-lymphocytes.

Cardiovascular system
- inhibits thrombosis.
- vasodilator effect.
- stimulates cyclic AMP formation and hence inhibits cholesterol synthesis.

Skin condition
- controls sebum production.

Metabolism
- has insulin-like actions and can potentiate the effects of insulin.
- may protect against abnormal cell proliferation.

Anti-inflammatory effect
- inhibits the synthesis of pro-inflammatory substances.
- blocks release of lysosomal enzymes, which are thought to be responsible for much of the damage caused during the inflammatory response.

Reproductive system
- may regulate effects of oestrogen, progesterone and prolactin in the luteal phase of the menstrual cycle.

Figure 4. Prostaglandin synthesis from the linoleic acid family.

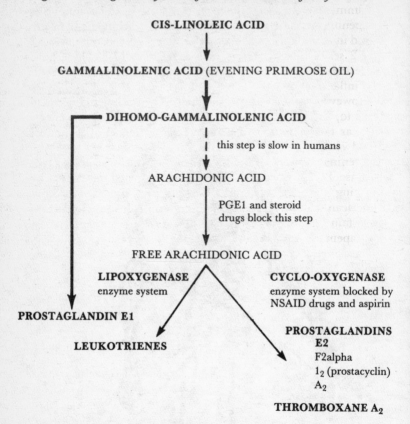

Leukotrienes, thromboxane and prostacyclin

Like prostaglandins, leukotrienes have an almost incredible variety of effects, some highly desirable and some harmful. Each cell produces a specific pattern of prostaglandins and leukotrienes. There are over 50 types of prostaglandins and leukotrienes and related molecules, and there are new ones being discovered each year.

Leukotrienes are not part of the picture of the conversion of evening primrose oil to its metabolites because dihomo-gammalinolenic acid (step 3 in the conversion process) cannot give rise to leukotrienes. However, evening primrose oil does have a role to play against the harmful leukotrienes.

Arachidonic acid gives rise to leukotrienes which are very inflammatory. In fact, whenever any kind of inflammation is happening in the body, the products of arachidonic acid are found in abundance. These other products are thromboxane A2 and 2 series prostaglandins such as PG F2alpha. These have undesirable effects such as promotion of vasospasm, thrombosis and inflammation.

However, some of the prostaglandins which arachidonic acid leads to, such as prostacyclin ($PG1_2$), do have desirable effects such as the inhibition of platelet aggregation and dilation of blood vessels.

Evening primrose oil works as an anti-inflammatory agent because PGE1 works in a similar way to steroid drugs, by blocking the mobilization of arachidonic acid [8,9]. Recently, American scientists have also found that it blocks the formation of certain harmful leukotrienes from arachidonic acid. Fish oils, eicosapentaenoic acid and its derivatives, seem to compete with arachidonic acid and so prevent the conversion of arachidonic acid to inflammatory metabolites. [10]

5 PREMENSTRUAL SYNDROME

Evening primrose oil has been used very successfully as part of a treatment programme for premenstrual syndrome (PMS) since the beginning of the 1980s. Trials have consistently proved that most women — more than 80% of them — who suffer from PMS improve on evening primrose oil.

Today, few people challenge the view that PMS is a condition which affects the whole system. It is no longer enough to call it PMT — premenstrual tension — as the symptoms are much more pervasive than tension alone.

Six of the most common symptoms experienced by women with PMS are irritability, depression, breast pain, bloating, headaches, and clumsiness.

However, the cluster of symptoms can include swollen ankles, legs and sometimes fingers, reduced libido, constipation, hot flushes, backache, nausea, acne, cramps, food cravings, lethargy and fatigue on the physical side, and, on the psychological and emotional side, anxiety, mood swings, suicidal impulses, low self-esteem, weeping for no obvious reason, sudden tantrums, lack of concentration and lapses of memory.

A woman going through PMS can feel fat and ugly, with an evil temper and a feeling that life is not worth living. PMS can cause havoc in a woman's life, at worst wrecking relationships, marriages, and careers. Women who suffer from PMS badly can be hell to live with and make unpleasant and unpredictable workmates. Women like this find themselves insufferable too while they are in the throes of PMS. Husbands and children often bear the brunt of their Jekyll and Hyde personalities.

As many as 40% of women aged between 15 and 50 get PMS symptoms of varying degrees, and about 10% gets them very badly indeed. At some time in their lives, about 80% of women experience some PMS symptoms.

For these symptoms to be true PMS, they typically happen up to 14 days before the onset of a period, and disappear when the period starts. The symptoms are similar every month.

PMS sufferers are most frequently in their thirties, and may be women who have had problems on the contraceptive pill, had hypertension during pregnancy, suffered from post-natal depression, and experienced periods of stress. Often, this group of symptoms goes hand in hand with PMS.

For years, women complaining of the emotional and psychological symptoms of PMS have either been told it's a woman's problem and is something they have to learn to live with, or else been given tranquillizers or anti-depressants, or seen a psychiatrist. The physical symptoms have usually been treated with hormones and diuretics. All these things have worked to some degree in some people, but have not got to the root of the problem for most women.

The factors which cause PMS

There are a great many factors which are thought to contribute to PMS.

- Lack of essential fatty acids and prostaglandin E1 (see below)
- Poor diet (too much tea, coffee, chocolate, sugar and dairy produce are common causes)
- Drink, drugs, and smoking
- (Possibly) environmental pollutants, eg petrol fumes, formaldehyde in furniture, chemical sprays
- The oral contraceptive pill
- Gynaecological and hormonal problems
- Stress and problems in life
- Lack of exercise
- Pregnancies (PMS symptoms sometimes get worse after a second or third pregnancy)
- Operations (PMS symptoms can occur after surgery such as a hysterectomy or sterilization, or abdominal surgery)
- Alcohol
- Candida albicans, the yeast infection, and food allergies. Candida albicans can either exacerbate or be mistaken for PMS. Food allergies are sometimes connected with PMS, and

can make the symptoms worse. Often Candida albicans and food allergies go together.

Why evening primrose oil is used in the treatment of PMS

Women suffering from PMS are thought to be low in essential fatty acids,[1,9] and also therefore to be low in the important prostaglandin made from EFAs, Prostaglandin E1. They also have an imbalance of the various prostaglandins.

A shortage of EFAs can lead to an apparent excess of the female hormone prolactin. Prolactin produces changes in mood and fluid metabolism, similar to those found in PMS. Body tissues may be abnormally sensitive to normal levels of prolactin when they are low in EFAs and PGE1.[2]

It is thought that GLA and PGE1, derived from evening primrose oil, can damp down these effects of prolactin and also of the hormones from the ovaries. The net effect of this is that GLA and PGE1 help smooth out the actions of the rapidly-changing hormone levels in the second half of a woman's menstrual cycle.

This is what two doctors running a PMS clinic in Massachusetts, USA, Ann Nazzaro and Donald Lombard, have said:

> Our clinical research experience has shown that when there is an increase in Prostaglandin E1 — that is, when we are provided the means to produce Prostaglandin E1 — there is a significant reduction in premenstrual syndrome problems. With a nutritional approach alone, we have found a 70% success rate...[3]

Treatment for PMS involving evening primrose oil

As PMS is such a pervasive condition, treatment has to involve many different things. Evening primrose oil is an important part of the nutritional therapy of PMS, but other things are important too, such as overall diet, specific vitamin and mineral supplements, exercise, and lifestyle changes to reduce stress. (See Maryon Stewart's *Beat PMT through Diet*.[4])

Vitamin, mineral and other supplements

At PMS clinics, women are prescribed a variety of vitamin,

mineral and other supplements to be taken with evening primrose oil.

At the Nazzaro and Lombard PMS clinic in the USA, they recommend the EFA co-factors of:

- Vitamin C: between 500mg to 3g a day
- B-complex tablet in appropriate ratios
- Vitamin B6: 50mg a day
- Zinc: 10mg a day

Vitamin B6. This vitamin has a particularly important role to play in the treatment of PMS. It is one of the co-factors in the metabolic conversion process of essential fatty acids. Some women may have below average levels of B6, and this is thought to have a harmful effect on the control centres of the menstrual cycle, the hypothalamus and pituitary glands. This vitamin is needed for the delicate inter-relationship between these two centres. Without it, they don't work properly. The contraceptive pill, with its synthetic hormones, can lower some women's levels of B6. Hormones may produce a condition where far more B6 is needed than usual — as high as 150mg a day. Doses much above this can lead to side-effects and should be taken only under medical supervision.

Overall, it seems that Vitamin B6 is helpful to any women with PMS, and also to women who get depressed when taking the contraceptive pill.[5]

In some circumstances, the additional nutrients of magnesium, chromiun, Vitamin E, and tryptophan are also used.

These nutritional supplements are part of a wider nutritional programme which involves cutting out all xanthines in the diet, found in substances like coffee, tea, chocolate and cola, as well as all trans fatty acids found in things like hard margarines and processed pastries.*

Other doctors go much further than this in their dietary recommendations and advise a nutritional plan which:

- Reduces intake of sugar and junk foods
- Reduces intake of salt
- Reduces intake of tea and coffee

*For full details of this treatment programme, please read *The PMT Solution* by Drs Ann Nazzaro and Donald Lombard with Dr David Horrobin, £4.95 from the Adamantine Press.

- Advocates eating green vegetables or a salad daily
- Limits intake of dairy products
- Reduces intake of tobacco and alcohol
- Advocates use of good vegetable oils
- Advocates plenty of wholefoods

In addition, they advise taking regular exercise.

These are the basics of the programme being used successfully by the Women's Nutritional Advisory Service in Hove, Sussex, which is explained in Maryon Stewart's *Beat PMT through Diet*.

The Women's Nutritional Advisory Service also uses evening primrose oil, plus vitamins and minerals, as part of its therapy. They carried out their own clinical (but unscientific) trials on products promoted for PMT in order to be able to give advice to their consumers as to which products really worked. Their results are shown in Figure 5.

Figure 5. Results of clinical trials performed by the Women's Nutritional Advisory Service.

Name and Type of product	Available From	Dosage Given	Improved a lot	Improved slightly	No worse but no better
Optivite	Nature's Best	4 per day increasing to 8 a day pre-menstrually	71.5%	16.5%	12%
Optivite	Nature's Best	2 per day increasing to 4 a day pre-menstrually	63%	31%	6%
Evening primrose oil (*Efamol*)	health food shops and chemists	8500mg capsules per day	64%	19%	17%
Premence 28	chemists	1 per day	72.1%	18.6%	9.3%

Note: These studies were performed on women who did not change their diet. The benefits would probably have been greater if all the nutritional guidelines had been followed at the same time. (None of these was a controlled study in that no placebo was tested.)

If you add together the 'improved a lot' and 'improved slightly' categories, a total of 83% improved on evening primrose oil.

These figures are similar to the findings of Dr Caroline Shreeve who has been using evening primrose oil to treat her PMS patients for several years.[6]

She says that every patient on evening primrose oil has gained some relief, and more than 90% have had total relief.

Scientific studies on evening primrose oil and PMS

The results of trials have been good, with upwards of 60% of women with PMS improving on evening primrose oil.

The most recent scientific trial (1985 – 1987) on evening primrose oil and PMS was at the Royal Free Hospital and St Thomas's Hospital in London. There has also been a recent study on PMS at the Universities of Helsinki and Oulu in Finland.[10]

The Royal Free and St Thomas's trial involved 60 women with PMS in a double-blind, placebo-controlled trial,[7] which assessed the effects of evening primrose oil compared with a placebo on the six common symptoms of PMS — depression, irritability, breast pain, clumsiness, bloating, and headache.

The results showed that all the six symptoms improved considerably. Overall, 60% improved on evening primrose oil (*Efamol*), compared with only 30% who took the placebo.

The Royal Free trial did not include any vitamins, minerals, dietary or lifestyle recommendations. The response rate is likely to increase if these are included.

In Sweden, evening primrose oil has had very good results on a wide range of PMS symptoms. Dr Bertil Larsson and Dr Stefan Fianu of Hudding University Hospital tried evening primrose oil (*Gammaoil Premium*) on 19 women aged between 25 and 48. These were their results:

Symptoms	Improvement
Swollen abdomen	95%
Breast discomfort	95%
Irritability	80%
Swollen fingers and ankles	79%
Anxiety	53%

However, fewer than half the women reported any benefits from evening primrose oil on the symptoms of tiredness and head-aches.

During 1981, a study was done at one of the major PMS clinics in the UK, at St Thomas's Hospital in London.[8] In this study, 65 women with bad PMS were treated with evening primrose oil (*Efamol*). All of them had tried one or more other standard treatment, and all of them had failed.

The results were good, with 61% of the women experiencing complete relief of their symptoms, and 23% partial relief. The other 15% said there had been no change as a result of taking evening primrose oil.

One particular symptom — breast discomfort — was helped considerably, with 72% saying this had improved. Other common symptoms that showed improvement after taking evening primrose oil were mood swings, anxiety, irritability, headaches, and fluid retention.

In the Finnish study it was found that Efamol particularly helped in relieving depression in the pre-menstrual syndrome.[10]

Dr Michael Brush, the biochemist who did the St Thomas's study, liked the idea of using a natural product with a nutritional approach as many of the drugs for PMS can have undesirable side effects.

In almost all cases, the starting dose was two capsules twice a day after food. Although a few patients were given treatment all through their menstrual cycle, most of them started treatment three days before the symptoms were expected to arrive and carried on until the start of their period. In a few very severe cases, the dose was increased to three capsules twice a day. Some of the patients were given Vitamin B6 (pyridoxine) at the same time.

A note about doses and side-effects

Doctors who run PMS clinics do not use the same doses of evening primrose oil. The best dose for the individual woman may have to be found by simple trial and error to see what works.

The lowest dose is used by Drs Nazzaro and Lombard in the USA. Interestingly, when patients have shown side effects of lethargy, or headaches, they have *lowered* the dose of evening primrose oil. These doctors also found that evening primrose oil occasionally had the side-effects of making the menstrual cycle longer, from 28 days to 31 or 32 days or even longer. It could also make people forgetful, with a feeling of being 'out of it'.

Drs Nazzaro and Lombard were getting an overall 70% success rate on doses as low as one or two capsules of 500mg evening primrose oil a day.

The Women's Nutritional Advisory Service in Hove, Sussex, reached their results of 83% improved on evening primrose oil when women were taking eight capsules a day.

The St Thomas's study used three 500mg capsules twice a day only in bad cases. The usual dose was two capsules twice a day, after food. In the St Thomas's study, three patients complained of minor skin blemishes during treatment, and three had passing phases of excessive mood quietening.

Efamol PMP, especially designed for PMS sufferers, suggests a dose of two capsules of evening primrose oil 500mg twice a day, together with one tablet of *Efavite* twice a day (containing the vitamin and mineral co-factors). This should be taken for ten days leading up to the period.

Case histories from the St Thomas's study

Case 1. A 26-year-old domestic manager, who gave a clear history of PMS from ten to 14 days before each period. This had gone on for the last six years. The main symptoms were severe breast discomfort, irritability, tearfulness, and poor co-ordination and concentration. Various hormone treatments had been tried without success. B6, 75mg twice daily, had shown good results at first. The woman improved when she took *Efamol*, three capsules twice a day, plus six tablets of *Efavite* a day.

Case 2. A 40-year-old school helper. She had attended the PMS clinic for four years. She gave a history of irritability, anxiety, depression, poor co-ordination, loss of libido, and moderate fluid retention for two weeks before each period. She had tried a hormone treatment and an anti-depressant, both without success. But when she took *Efamol*, two capsules twice a day, plus one *Efavite* a day, plus 100mg of B6 a day, her symptoms disappeared.

Case 3. A 34-year-old school secretary. She had suffered from PMS since coming off the pill five years previously, and it had worsened after she had been sterilized three years later. She complained of severe breast pain, swelling of the stomach and face, irritability and depression. This happened for the last 14 days of a regular 28 to 29 day cycle. Tranquillizers, diuretics, and B6 had not helped. She was put on *Efamol*, four capsules a day, plus B6, 80mg a day. This improved her mood, the swelling and breast discomfort.

Case 4. A 34-year-old housewife. She had suffered from PMS since the birth of her first child eight years ago. The symptoms were lethargy and mood swings, irritability and depression, general bloating, very painful breasts and loss of co-ordination for 14 days before her period. Various drugs had not worked. *Efamol*, four capsules a day, plus *Efavite*, four tablets a day, have given her complete relief of her symptoms.

6 BENIGN BREAST DISEASE

Doctors running a mastalgia clinic in Cardiff, Wales, say that evening primrose oil is 'the most promising first-line treatment, especially for patients under 25, since side-effects are uncommon and it has good patient acceptability.

In benign breast disease — known medically as mastalgia — the breasts feel lumpy and tender, and they have a granular sort of texture. The breasts can be painful and swollen, with inflamed fibrous tissue and cysts. All these symptoms can be worse just before a period begins. In these cases, the condition is said to be cyclical. There are also women with non-cyclical breast disease which does not vary with the menstrual cycle. No one knows exactly what causes benign breast disease.

Many women who suffer from pain in their breasts may fear the worst and worry that they have breast cancer. So the most important thing to do first if you do have painful breasts is to see your doctor so he can send you for further investigations. These might include a physical examination, a mammography and perhaps a biopsy.

For women where cancer has been excluded, and who still suffer from unremitting breast pain, evening primrose oil is helpful in many cases.

Evening primrose oil and painful breasts

In a recent study, evening primrose oil was found to help 45% of women suffering from cyclical mastalgia, and 27% of women suffering from non-cyclical mastalgia.

The study was done at the University of Wales College of Medicine on 291 patients with severe persistent breast pain.[1]

Evening primrose oil was one of three drugs used in the study.

The others were danazol and bromocriptine. Although danazol had a success rate of 70% and bromocriptine 47% in cyclical mastalgia, both these drugs had side-effects in many women. That's why the doctors conducting the study came to the conclusion that it was worth trying evening primrose oil first as it was virtually free of side-effects and easy to take. If evening primrose oil didn't work, only then should danazol and bromocriptine be tried.

The Cardiff study showed that if one drug didn't work, another drug might. So which drug is right for which woman can only be found by a process of trial and error.

The dose of evening primrose oil used was six capsules of 500mg daily for three to six months. Ninety-five % of those responding to evening primrose oil did so within three months.

Many women suffered a relapse shortly after stopping the evening primrose oil. This happened within eight weeks after the end of the course of treatment. Most women found that for the best results they had to go on taking the evening primrose oil for long courses of treatment — a minimum of six months.

The overall results of the Cardiff study were that four out of five patients with cyclical mastalgia did gain relief of their symptoms on one of the drugs used.

The Cardiff study confirms the results of previous studies done in Cardiff and Dundee [2] which showed that evening primrose oil considerably helped women with benign breast disease, particularly when their symptoms were cyclical. These women found that the tenderness and lumpiness of their breasts was significantly reduced after three months on evening primrose oil (*Efamol*). For women who had non-cyclical symptoms, three months' treatment on evening primrose oil produced a marked improvement in their breast tenderness, but the lumpiness did not improve.

Painful breasts can be a symptom of PMS (see Chapter 5). In the original St Thomas's Hospital study, Dr Michael Brush found that six out of nine cases of severe breast discomfort with fibrous cysts responded to evening primrose oil treatment.

How evening primrose oil works in benign breast disease

Benign breast disease may have something to do with a high intake of saturated fat. There is an interesting relationship between heart disease in men, and benign breast disease and PMS in women. In societies where there is a high rate of death from heart disease among young and middle-aged men, there is a correspondingly high rate of breast disease and PMS in young and middle-aged women. So, wherever saturated fat intake is high relative to the intake of essential fatty acids, breast and menstrual cycle disorders become common.

If a high intake of saturated fat is associated with benign breast disease and other disorders, then increasing the intake of polyunsaturated fatty acids may reverse or prevent the development of benign breast disease and menstrual cycle problems.

Women with breast disease have high rates of sebum production, which is a marker of EFA deficiency.[3] Also, PGE1 inhibits some of the peripheral actions of prolactin, a hormone which has been implicated in both breast and menstrual cycle problems.[4]

A shortage of essential fatty acids in the diet does lead to excessive amounts of fibrous tissue. Cysts, which are another common symptom of benign breast disease, may form because the body is for some reason making too much of the hormone prolactin and is also short of Prostaglandin E1. As a group, women with benign breast disease do seem to have higher levels of prolactin in their blood than they should.

The idea behind giving evening primrose oil as a treatment for benign breast disease is that PGE1 can dampen down the effects of prolactin, may help prevent the development of cysts, and can help remove lumpiness in the breasts.

Diet

For best results, evening primrose oil should be taken in conjunction with a healthy, low saturated fat diet. In particular, the methylxanthine group of substances (caffeine, theophylline) found in coffee, tea and cola drinks should be excluded, as these substances increase the binding of prolactin to the breast. Removal of these stimulants from the diet has improved benign breast disease.

Breast size

An interesting, unexpected and usually welcome effect of evening primrose oil in some women is that it seems to make their breasts bigger. No one knows exactly why.

This phenomenon was first reported in 1981 at a symposium on evening primrose oil when several women revealed they had gone up several bra sizes since first taking evening primrose oil — but without putting on weight anywhere else. All the evidence so far has been anecdotal, and there have been no scientific studies to investigate this finding. It is worth noting that the women who have reported bigger breasts have usually been taking evening primrose oil for more than six months, and in some cases for years.

7 ECZEMA

As long ago as the 1930s it was known that essential fatty acids are vital for healthy skin and hair.

But it is only in the last ten years or so that essential fatty acids — evening primrose oil in particular — have proved to be an effective treatment for some common skin complaints, especially eczema.

What happens to the skin when there is a low level of essential fatty acids has been shown in animal studies. The main findings are:

- The skin becomes scaly and rough and sheds dandruff-like scales. In severe deficiency, an eczema-like dermatitis may develop and the skin may break down.
- Wounds take longer to heal.
- There is greater water loss from the skin, making it drier and making it age more quickly.

Atopic eczema

Atopic eczema is a chronic, patchy, mild inflammation of the surface of the skin which almost always begins in infancy or early childhood. It can be made worse by irritants, but often occurs without any apparent cause. The main signs and symptoms are dry, scaly, itchy red patches on the skin. Atopic eczema is different from contact eczema which is an inflammation caused by contact with an irritant, such as a chemical or a cosmetic.

In recent years, evening primrose oil has had excellent results in the treatment of atopic eczema, in both adults and children. In several trials done so far, the results of evening primrose oil on atopic eczema are good or very good in about three quarters of all patients.

Improvement does not happen overnight. The first signs of improvement in moderate or severe cases of atopic eczema will be in the first four weeks at the earliest, but almost certainly between four and 12 weeks after starting to take evening primrose oil.

Many people who have taken evening primrose oil to treat their eczema have carried on taking it after the end of the period of trial, sometimes for years.

Children with eczema

Atopic eczema can show in infants as young as only a few months old. It sometimes appears for the first time when a baby is weaned off breast milk.

Recent research done in Italy[1] and Sweden[2] shows that most children with eczema can improve considerably on evening primrose oil. Only a few years ago, it was thought that children with eczema did not respond as well as adults, but this may have been because they were not given a high enough dose. It is known that growing animals have far higher essential fatty acid requirements than adults and the same probably applies to humans. Now we know that children do in fact respond very well indeed.

In Bologna, Italy, Dr Alessandra Bordoni studied 24 children with atopic eczema. Twelve of them were treated with evening primrose oil and 12 with a placebo. The group treated with evening primrose oil (*Efamol*) significantly improved compared with the placebo group. 65% of the group taking evening primrose oil showed significant improvement after just four weeks of therapy. The dose was 3g a day (i.e. six capsules of 500mg).

It is not yet understood why some children did not respond. There is also a small percentage of adults with atopic eczema who do not respond to evening primrose oil treatment. Perhaps this is because there is no essential fatty acid abnormality in these cases.

Evening primrose oil can safely be given to babies and very young children. If they are too young to take the capsules by mouth, the oil can be taken out of the capsules by pricking the gelatin shell. The oil can then be rubbed into the soft parts of the skin, like inside the thighs and on the tummy. The oil penetrates the skin very quickly. Contrary to what you might expect, you should rub the oil into the healthy skin and not the areas affected by eczema. However, for children old enough, it is better to take the evening primrose oil by mouth as the conversion process is more efficient.

Breastfeeding is able to protect against the development of atopic conditions, including atopic eczema.[3,4] Human breast milk, unlike cow's milk, is rich in gammalinolenic acid, dihomo-gammalinolenic acid, and arachidonic acid. As a matter of interest, a six-month-old, fully breast-fed baby is getting the equivalent amount of GLA to about three 500mg capsules of evening primrose oil a day.

Adults with eczema

More and more studies from around the world are confirming that evening primrose oil can help adults with atopic eczema. The first trial, at Bristol Royal Infirmary,[5] showed that evening primrose oil worked best in higher doses, and when it had been taken for several months. The most effective dose was eight to 12 capsules of 500mg (4 to 6g) evening primrose oil taken for longer than 12 weeks.

Other trials on adults with eczema during the 1980s have tended to confirm that large oral doses (4 or 6g a day) of evening primrose oil produce the best results. They also confirm that the longer the patient takes evening primrose oil, the better the results.

In a multi-centre study involving 13 hospitals in England, Scotland, Denmark and Finland,[6] the doctors found that evening primrose oil produced excellent results in a very difficult group of patients, who all had a prolonged history of eczema, mostly starting in infancy or early childhood. On average, the patients had suffered from eczema for 17 years without having found any effective treatment.

Nearly all the patients reported improvement from the evening primrose oil treatment in every symptom of eczema — redness, dryness, scaling, itch, and swelling. Fifty-three out of 54 patients suffering from moderately severe eczema and 58 out of 59 patients with severe eczema reported an improvement in their overall condition.

During the treatment with *Efamol* many patients were able to reduce or stop altogether their use of antihistamines, antibiotics, systemic steroids and the most potent grade of topical steroids. After treatment with evening primrose oil, some patients could halve their doses of potent and moderate topical steroids, and could cut down on their use of antihistamines by a quarter, and of antibiotics by about a sixth. Apart from the use of emollients, they could dramatically reduce all their usual medication.

Nearly all the 116 patients (71%) in this multi-centre study chose to keep on taking *Efamol*. For some, this has run into several years. The average time for taking it was 11 months. Fifteen patients who stopped taking *Efamol* noted a relapse of their eczema.

A double-blind trial in Finland [7] in 1987 showed that evening primrose oil had a significant effect on patients with atopic eczema. The overall severity was reduced, there was less inflammation, there was a reduction in the percentage of the body surface covered by eczema, and the patients suffered from less dryness and less itching. The dose in this trial was eight capsules (four capsules twice a day) for 12 weeks.

In this trial, patients were allowed to use emollient creams at the same time if they wanted to. They were also allowed to use a mild topical corticosteroid cream or oral antihistamine, or both, during the trial if they had severe skin symptoms. Interestingly, three times as much topical steroid was used by patients in the placebo group than in the group taking evening primrose oil. This suggests that the evening primrose oil was working much better than the placebo.

The doctors conducting the trial came to the conclusion that evening primrose oil was an adjunct to the existing treatment for atopic eczema, and could be safely used alongside steroids and other creams.

Evening primrose oil as a treatment works best in cases of moderate or severe atopic eczema. It has not been found to work well in cases of very mild eczema, where there is no family history of atopic disorders, and where the essential fatty acid levels in the blood were normal to start with.

Pets with eczema

As a result of some studies [12,13,14] done on dogs with allergic eczema, some vets are now recommending evening primrose oil for dogs with bad skin conditions.

The Royal Veterinary College conducted studies [12] with evening primrose oil, and also with evening primrose oil with added fish oils, on 16 atopic dogs with dermatitis.

This is a summary of their results:

Treatment with evening primrose oil.

Breed	Total	Responded
Border collie	1	1
Boxer	1	1
English setter	4	1
German shepherd	2	2
Poodle	3	2
Springer spaniel	1	1
Staffordshire bull terrier	1	1
West Highland white	3	1
	16	10

The symptoms studied were itch, dryness, coat condition, scale, erythema, and oedema. All the symptoms except itch responded well. There was new hair growth in previously affected areas of the body, which were now healing.

In the second stage of the study, twelve dogs were also given fish oils. However, no great further improvements were seen, possibly because many of the dogs had reached such an improved state at the end of stage 1.

Since these studies, another Efamol formulation has been given to some dogs with encouraging results. The formulation, called *Efaderm*, combines evening primrose oil with its vitamin and mineral co-factors.

How evening primrose oil helps

Patients with atopic eczema show some of the signs of essential fatty acid deficiency. But in fact their blood has been found to have above-average levels of linoleic acid and also of alpha-linolenic acid. So the problem is not that people with atopic eczema are eating too little of the foods containing the parent essential fatty acids. Rather, there seems to be some problem in using these fatty acids. They are not being metabolized properly.

The blood of someone with atopic eczema is typically very low in the metabolites of linoleic acid and alpha-linolenic acid, which indicates that there is probably an enzyme block stopping the conversion of these essential fatty acids.

All the studies done so far agree that people with atopic eczema have below-normal levels of GLA, DGLA, AA, PGE1 and the metabolites of alpha-linolenic acid. The enzyme delta-6-desaturase is needed to get from linoleic acid to the next step, and from alpha-linolenic acid to the next step.

Evening primrose oil completely by-passes this enzyme block by starting at the next stage in the metabolic pathway of the linoleic acid family. (It has no effect on the alpha-linolenic family. Fish oils should also be taken to help correct the low level of metabolites of the alpha-linolenic acid family.)

Various studies have been done to see what happens to the fatty acid profile of the blood after people with atopic eczema have been taking evening primrose oil. Overall, the results are that evening primrose oil can go some way to correcting this abnormal blood profile and make it more normal.

No one knows exactly why the delta-6-desaturase enzyme may be defective. There are many possible reasons for this, including a minor abnormality in the protein structure of the enzyme or an abnormality of co-factors.

Asthma, hay fever, allergies and other atopic conditions

On the face of it, eczema, asthma, hay fever and allergies all sound like very different conditions. But in fact they have a lot in common — they are all to do with an abnormal body defence system. Doctors call this condition 'atopy'.

In fact atopy — or a generalized allergic response — can show itself as any or all of a variety of conditions. As many as one in five of the population suffers from some sort of atopy (though this term is virtually unknown by the layman). Atopy is common in patients with ulcerative colitis, Crohn's disease, ear problems, nasal polyps, and some obstetric problems.

Atopic eczema is closely linked with other atopic conditions like asthma and hay fever, and it is common to find other members of the family suffering from these things. In some ways, atopic eczema behaves like a type of asthma where the patient is a little short-winded virtually all the time and occasionally has real difficulty in breathing. In one person the atopy shows up as eczema, but in another person it might take the form of, say, asthma.

There are several things in common between eczema, asthma, allergies and other atopic conditions:

1 Faulty immune response. It has been known for a long time that people with eczema, asthma and allergies have something wrong with their immune system. There is some speculation that the abnormalities of the immune system in atopic disease are partly secondary consequences of a disordered fatty acid metabolism. If there is a fatty acid abnormality, various parts of the immune system or things which regulate the immune system are badly affected, particularly PGE1 and the T-lymphocytes. The abnormal fatty acid composition found in people with atopic eczema has similarities with cases of respiratory allergy.

2 Faulty enzyme function. Atopic people may have a defect in the delta-6-desaturase enzyme, which is needed to convert linoleic acid to GLA. The fact that evening primrose oil works in atopic eczema means that the enzyme block can be by-passed, which would indicate that a defective enzyme is the guilty party. This may also be the case with other atopic conditions.

The blocking agents described on page 31 are inhibitors of the delta-6-desaturase enzyme. So people with atopic conditions must be more careful about the things which cause disruption to an already defective enzyme system.[8] The main ones are:

• Trans fatty acids
• Too much saturated fat
• Simple sugars
• Alcohol
• Catecholamines — hormones released by adrenal glands during stress

Evening primrose oil does nothing to correct the actual defective enzyme. But, by starting at step 2 in the conversion process of linoleic acid, it gives the body enough essential fatty acids for everything to be able to work properly.

So evening primrose oil helps correct the faulty immune system in people with atopic conditions. This is because it converts to PGE1, which stimulates the T-lymphocytes, which play a key role in the immune system. T-suppressor lympho-

cytes are a type of white blood cell which seem to keep other parts
of the immune system under control and which make sure that
the immune system first and foremost attacks foreign invaders,
like bacteria and viruses, and not the body's own tissues.

It seems that the T-lymphocytes, especially T-suppressor
cells, are faulty in people with atopic conditions.[9,10] When
T-suppressor cells are defective, auto-immune damage often
happens.

Cystic fibrosis

Children with cystic fibrosis seem to have a severe block of the
delta-6-desaturase enzyme — much more severe than in atopy.
This makes them extremely deficient both in EFAs and in
PGE1. They also have a problem absorbing fats. They do not
respond to treatment with safflower oil, which is rich in linoleic
acid.

Some people with an atopic condition are thought to be
carriers of cystic fibrosis. An atopic parent could be said to have
a single dose of abnormality in the enzyme block, whereas a
cystic fibrosis child has a double dose of the abnormality.[11]

8 HYPERACTIVE CHILDREN

Evening primrose oil has helped to improve dramatically the lives of countless children and their families. Together with other nutritional approaches, it can help turn a little monster into a normal, happy, loving child.

It seems that evening primrose oil works especially well on *atopic* children (see previous chapter) where there is a family history of such ailments as eczema, asthma, allergies, hay fever or migraine. The mothers of hyperactive children are often found to have migraine, and to have suffered from pre-menstrual tension and post-natal depression.

In a questionnaire conducted by the Hyperactive Children's Support Group (HACSG) in February 1987, a total of 92% of the children came from an atopic family. Of these 102 children, 34% suffered from either both or one of eczema and asthma.

Not all hyperactive children respond to evening primrose oil. One sign that they may do so is excessive thirst. In the HACSG study 78% of the children did have abnormal thirst. Unlike in diabetes, this need to drink a lot is not accompanied by excessive urine. Instead, the urine is rather concentrated which probably means that these children are losing water through the skin. Excessive thirst and permeability of the skin are signs of essential fatty acid deficiency.[1]

Diet

Evening primrose oil is by no means the first or the only treatment for hyperactive children. The first thing to do is to change their diet. Ninety-four % of hyperactive children are helped by diet.

The most successful nutritional approach to hyperactivity in children is the Feingold diet, so named after the late Dr Ben Feingold, the American doctor who invented it. The HACSG recommend a diet based on Dr Feingold's but with a greater restriction on additives.

The key things to take out of the child's diet are artificial colourings, flavourings and preservatives, and naturally-occuring salicylates found in such seemingly innocent foods as apples, oranges, peaches, strawberries, grapes, cherries, almonds and cucumbers. Salicylates in drugs such as aspirin are also excluded.

Salicylates are known to block the formation of prostaglandins. Hyperactive children might be deficient in PGE1, which helps control the immune system, and has an influence on such things as asthma, behaviour, and thirst (via the kidneys). Low levels of PGE1 can be corrected by taking evening primrose oil.

Hyperactive children are also very often sensitive to chemicals such as those in glue, felt-tip pens, aerosol cleaning fluids and so on.

Parents are advised to give their child only fresh foods and to avoid anything with 'E' additives, plus anything containing salicylates. If the child's behaviour is obviously made worse by certain chemicals, then it also means switching to other things which do the same job, such as using wax crayons instead of felt-tip pens.

Food allergies

In many cases, a change in diet alone will be sufficient to see a huge improvement in a hyperactive child. Generally speaking, the younger the child, the quicker and better the response.

However, some children also have some specific food allergies. In many cases, cow's milk and all cow's milk products can be a problem for hyperactive children.

The only sure way to find out which foods a child is allergic to is to have him (most are boys) tested for a whole range of foods, or to cut out suspect foods from his diet.

Vitamin and mineral supplements

The HACSG has found that giving hyperactive children vitamin and mineral supplements is very helpful, in addition to making changes in their diet. In many cases they also suggest a hair analysis should be done in order to detect deficiencies of trace elements and minerals.

Almost all hyperactive children who are tested in this way are found to be very low in zinc and magnesium, although other anomalies do sometimes also show up. For example, some children have been found to have abnormal amounts of aluminium and lead, which could be responsible for some aspects of their behaviour or learning difficulties.

A general vitamin/mineral supplement PLUS a special zinc/B6/magnesium supplement has proved to be very helpful in many cases. Zinc and magnesium are both needed to metabolize essential fatty acids properly.

Evening primrose oil and hyperactive children

Originally it was thought that all the symptoms of hyperactivity could be put down to a lack of essential fatty acids.[1] Now it seems that only some of them can be laid at the door of an EFA deficiency.[2,3] We are now sure that evening primrose oil works best for those hyperactive children who are also atopic or who come from an atopic family.

There is now a great deal of evidence that people who suffer from an atopic condition, such as eczema, are not utilizing essential fatty acids properly. They may be eating enough in their diet, but it is not getting through. Like other atopic people, hyperactive children might have something wrong with the delta-6-desaturase enzyme, which is needed to convert linoleic acid to the next step in its metabolic pathway. Evening primrose oil avoids the enzyme block because it starts at the next step (see Figure 6).

Evening primrose oil should always be taken with its co-factors. These are zinc, Vitamin B6, nicotinamide (Vitamin B3), and Vitamin C. These co-factors are essential if the evening primrose oil is to work properly. The oil will not work so well without these co-factors. The general vitamin and mineral supplement is also sensible so that all the vitamins, minerals and trace elements can work in a proper balance with each other.

When evening primrose oil is taken together with its co-factors, the effects can be startlingly good. The result is calmness, concentration, a change of outlook, and improved general health.

Figure 6. Diagram to show how the metabolism of essential fatty acids can be hampered or helped in hyperactive children.

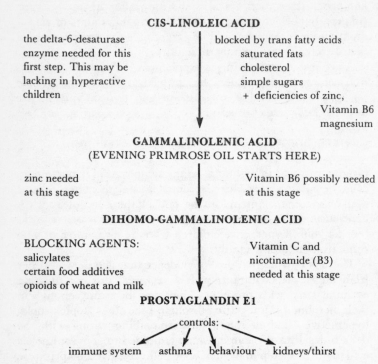

CIS-LINOLEIC ACID

the delta-6-desaturase enzyme needed for this first step. This may be lacking in hyperactive children

blocked by trans fatty acids
saturated fats
cholesterol
simple sugars
+ deficiencies of zinc,
Vitamin B6
magnesium

GAMMALINOLENIC ACID
(EVENING PRIMROSE OIL STARTS HERE)

zinc needed at this stage

Vitamin B6 possibly needed at this stage

DIHOMO-GAMMALINOLENIC ACID

BLOCKING AGENTS:
salicylates
certain food additives
opioids of wheat and milk

Vitamin C and nicotinamide (B3) needed at this stage

PROSTAGLANDIN E1

controls:

immune system　asthma　behaviour　kidneys/thirst

Dose

Age under two years: please contact HACSG (see the address at the end of the book) as it is necessary to work out individual requirements carefully.

Age two to five years: two evening primrose oil capsules of 500mg a day, rubbed into the skin or taken by mouth. Plus multi vitamin and mineral tablet, plus co-factors: two tablets per day Vitamin C 250mg, nicotinamide 15mg, Vitamin B6 50mg, zinc sulphate 5mg.

Age six to seven years: three evening primrose oil capsules of 500mg a day, either taken by mouth or rubbed into the skin.

Plus a multi vitamin/mineral tablet, plus co-factors: three tablets per day of Vitamin C 375mg, nicotinamide 22.5mg, Vitamin B6 75mg, zinc sulphate 7.5mg.

Age seven upwards: four evening primrose oil capsules of 500mg per day, either taken by mouth or rubbed into the skin. Plus a multi vitamin/mineral tablet, plus co-factors: four tablets Vitamin C 500mg, nicotinamide 30mg, Vitamin B6 100mg, zinc sulphate 10mg. Dose can gradually be increased to six evening primrose oil capsules. From the ages of eight or nine upwards, five or six capsules of 500mg evening primrose oil are effective. Begin with small amounts and gradually increase.

For each age group, take the tablets twice a day, once in the morning and once in the evening with food. If you rub the oil into the skin rather than have the child take it by mouth, choose a soft part of the body such as the abdomen, top of thighs, or inside forearms. The oil is very fine and can be rubbed in until it is completely absorbed. The gelatin shell of the capsules can easily be punctured by a safety pin, and the oil can then be squeezed out into the palm of your hand.

Case histories

Some very moving case histories come from the files of the Hyperactive Children's Support Group.[4]

'Anthony's improved with evening primrose oil. If this is so effective, why oh why can't doctors use it?

'Anthony sat on my knee and watched television, cuddled up. This is the first time since he was born and he is four in September....'

'Anastasja is greatly changed: aged $5\frac{1}{4}$ when she started *Efamol* etc. She learned to read and swim, to tie her laces, began judo, ballet and gymnastics and wets the bed far less often. She is a far happier child now.'

'There has indubitably been a dramatic improvement in Gerald's school results, such improvement coinciding exactly with the commencement of the treatment (supplements). The school assesses children fortnightly on a scale which ranges from -7 to $+7$ for the total work. Prior to the treatment Gerald had never achieved a mark above 0 and was normally around -4 to -5. After starting the treatment his first assessment was $+2$ and subsequent assessments have been $+3$ or $+3\frac{1}{2}$. He has even for the first time been picked for school sports teams and a solo in a musical concert. Gerald received a prize for the "most improved" child in the school this term.'

Jonathan is the son of a single mother, who ran out of evening primrose oil partly due to the cost. This is her story of what happened when Jonathan stopped taking his supplements:

'For those 10 days (when he was without his oil etc.) I had noticed that he had dark circles under his eyes, the old white complexion, hyperactivity, stupid, cheeky activity, distress, fighting for nothing — tears — and on Saturday ALL DAY, I lived through the nightmare which I haven't experienced for 2 years or more. I got it all. He ran away twice, total non-compliance, "he is going to burn the house down" hysteria, crying, tormenting me — all day....'

This mother then put her son back on evening primrose oil and the other supplements. Her letter goes on:

'Jonathan is manageable again for the moment as I have now increased his dose — but it will take time again....'

Mother of four children Nikolette Bennett wrote this success story about her hyperactive son Christopher in *Alive* magazine in Canada:

'At the end of a particularly disastrous day, I decided to try *Efamol* of which I had read positive reports. I began by rubbing a capsule of oil onto his wrists, every day. Within a week, a truly remarkable change took place. Christopher's speech modulated, the door ceased slamming, and for the first time, he sat through and ate up all of his dinner. He stopped demanding dessert, and ate his breakfast cereal without sugar!

'I find that I don't have to use *Efamol* every day, now. In fact, Christopher knows himself when he needs it. His voracious appetite for sugared foods has disappeared....

'The most wonderful aspect of being able to meet and overcome the challenge of Christopher's hyperactivity is that at last we are able to express positive love for our delightful son. Christopher is happier within himself and about himself and our family lives in harmony again.'

9 RHEUMATOID ARTHRITIS

The most exciting recent development with evening primrose oil has been with rheumatoid arthritis. Evening primrose oil may help many patients reduce their dose of non-steroidal anti-inflammatory drugs, or give them up altogether.

A study was done at Glasgow Royal Infirmary with 49 patients with rheumatoid arthritis.[1] Sixteen of these patients were given evening primrose oil (*Efamol*), 15 were given a combination of evening primrose oil and fish oil (*Efamol Marine*), and 18 patients a placebo (liquid paraffin.)

The aim of the study was to find out whether evening primrose oil or evening primrose oil combined with fish oil could replace the conventional non-steroidal anti-inflammatory drug (NSAID) treatment in rheumatoid arthritis.

The study lasted 15 months altogether. The initial 12 month treatment period was followed by three months of placebo for all groups. The dose was 12 capsules, usually taken as 3 capsules four times a day. No one knew what they were taking as all the capsules looked identical.

At the end of 12 months, results showed a significant subjective improvement for the *Efamol* and the *Efamol Marine* group, compared with the placebo group. Moreover, by 12 months the patients on *Efamol* and *Efamol Marine* had significantly reduced their NSAIDs. And despite this decrease in drugs, the disease did not get worse.

Sixty % of the patients on the *Efamol* alone were able to stop taking drugs, and a further 25% were able to halve the dose without ill effects.

The most consistent results, though, were for the *Efamol Marine* group. In this group, 60% were able to stop taking their anti-inflammatory drugs altogether, while 35% halved their dose. However, the amount of subjective improvement in the *Efamol* alone group was often greater than in the *Efamol Marine* group.

It was clear from this study that the evening primrose oil, or combination of oils, was the therapeutic anti-inflammatory agent. When the treatment group was switched to placebo for the last three months of the trial, all but one of the patients suffered a relapse.

The conclusion of this important study is that evening primrose oil and evening primrose oil with fish oils produce a subjective improvement in rheumatoid arthritis, and allow some patients to reduce or stop treatment with conventional anti-inflammatory drugs.

However, as yet there is no evidence that they act as agents which actually modify the disease.

The patients who had been taking evening primrose oil or evening primrose oil combined with fish oils experienced sub-jective improvements in their condition, and felt a greater sense of well-being. However, the doctors working on this study were not able to measure any objective improvements.

All the patients in the Glasgow trial had a fairly mild form of rheumatoid arthritis. The results of this particular trial were much better than some previous ones in which a lower dose of evening primrose oil was given for a shorter period of time on more severely disabled patients.

The doctors conducting this study felt that, as those patients on the *Efamol* and the *Efamol Marine* treatment were able to decrease or stop their usual drugs, these oils can best be used in those cases where a patient cannot take NSAIDs because of conditions such as a peptic ulcer or renal impairment.

The great advantage of evening primrose oil for rheumatoid arthritis is that it is a natural product, without side-effects. In contrast to the NSAIDs, evening primrose oil has actually been shown to have a protective effect on the stomach.

Dr Jill Belch, senior registrar in rheumatology at Glasgow Royal Infirmary, said after the results of the study were published that 'these results clearly show that this line of treatment is promising'.

Rheumatoid arthritis, inflammation, and drugs

Rheumatoid arthritis is the inflammatory form of arthritis. It is a chronic disease affecting connective tissue, mainly of joints, and can be very painful. It is thought to be an auto-immune disease, where the immune system attacks its own body, instead of defending it. The disease affects about one person in 20 in Britain at some time, and the figure rises dramatically after the age of 65. Some people are seriously crippled by the condition.

Put very simply, the problem with rheumatoid arthritis is that the body is producing too many of the wrong kind of prostaglandins, which have an inflammatory effect. These are the 2 series prostaglandins.

There are plenty of 2 series prostaglandins in people with rheumatoid arthritis, but not enough of the 1 series.[2] The 2 series PGs are made from arachidonic acid in the diet, especially meat and dairy produce, whereas the 1 series PGs are made only from essential fatty acids. If there are not enough essential fatty acids, either because not enough are eaten or not enough are getting through the system, then not enough 1 series prostaglandins will be manufactured.

When PGE1 is in short supply, the immune system is one of the first things to feel the pinch. PGE1 has a very important job in controlling T-lymphocyte production,[3] and the T-lymphocytes are the crack troops of the body. When the front line is absent, mayhem ensues.

Modern drugs used to treat arthritis, steroids and non-steroidal anti-inflammatory drugs (NSAIDs) work by inhibiting *all* prostaglandins, both the good and the bad, and in the case of steroids, other substances called leukotrienes, which are thought to cause inflammation.

Steroids inhibit the mobilization of arachidonic acid from phospholipids, and so reduce the formation of both leukotrienes and prostaglandins.

Most NSAIDs reduce the formation of prostaglandins by blocking cyclo-oxygenase, which gives rise to prostaglandins.

But the trouble with conventional drug treatment is that the drugs not only suppress inflammatory substances, but they also suppress anti-inflammatory ones as well. They knock out *all* prostaglandins, the helpful PGE1s as well as the unhelpful PGE2s.

Figure 7. How conventional drugs work in rheumatoid arthritis.

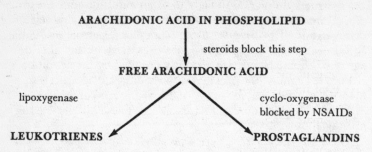

Evening primrose oil and rheumatoid arthritis

In rheumatoid arthritis, the prostaglandins are out of balance. There are too many of the pro-inflammatory PGE2s, and not enough of the anti-inflammatory PGE1s. No one yet knows exactly why this is. Perhaps because not enough foods with cis-linoleic acid are being eaten. Perhaps one or more of the blocking agents is getting in the way of the conversion process. Perhaps there's a shortage of any or all of the important co-factors, Vitamin C, Vitamin B6, B3, zinc and magnesium.

Whatever the reason, evening primrose oil avoids all the hurdles and pitfalls. It is able to start up the manufacture of PGE1 because its active ingredient — GLA — is a precursor of PGE1. Once PGE1 is on the production line again, the normal balance between the 1 and 2 series PGs can be restored. This is because when PGE1 levels go up, the PG2 levels go down, like a see-saw. Also, evening primrose oil reduces the production of harmful leukotrienes.

Taking evening primrose oil is a more physiological approach to the treatment of rheumatoid arthritis than taking drugs, in that it allows the body to make its own PGE1.

There are several ways in which evening primrose oil might work in rheumatoid arthritis. However, some of these can only be described in rather technical language.

1. The human thymus (which produces T-lymphocytes) and lymphocytes are relatively rich in PGE1.[4] PGE1 has several actions which are similar to those of thymic hormone and seems to be necessary for the normal working of T-lymphocytes.[5,6] T-lymphocytes are known to be abnormal in rheumatoid arthritis and other inflammatory disorders.

2. Lysosomes are organelles in the cells which contain a variety of destructive enzymes. These enzymes can be released during inflammation and may be the cause of much of the damage in rheumatoid arthritis. PGE1 can block the release of lysosomal enzymes.

3. Arachidonic acid is released from its cellular stores in most types of inflammation. It is converted to 2 series PGs and to substances called leukotrienes, both of which are powerful inflammatory agents. Evening primrose oil, by converting to PGE1, reduces the conversion of arachidonic acid to 2 series PGs and to leukotrienes.[10]

The study on human patients in Glasgow has confirmed animal studies using evening primrose oil for arthritis. Dr Robert Zurier and his group from Pennsylvania pioneered the idea that PGE1 is anti-inflammatory. He and his researchers showed that PGE1 can inhibit experimental ('adjuvant') arthritis in rats; control the systemic-lupus-like syndrome in NZB/W mice; activate T-lymphocytes; and control lysosomal enzyme release in humans.[7,8]

Zurier's group showed that evening primrose oil was as effective as PGE1 in controlling experimental arthritis in rats.[9] (In the first experiment, PGE1 itself was injected. In the second experiment, evening primrose oil was used, which converted into PGE1 inside the rat's body.)

The mechanisms by which evening primrose oil works in rheumatoid arthritis are likely to be the same for other inflammatory disorders.

10 DIABETES

Even when diabetes is well controlled by insulin and diet, there can still be complications. Diabetes can lead to severe damage to the heart and circulation, to the eyes, to the kidneys, and to nerves. At least half of all patients with diabetes develop one or more of these complications in spite of the best current treatment.

The damage to the nerves — known medically as diabetic neuropathy — can lead to loss of skin sensation, skin problems, muscle weakness, bladder and intestinal problems, and impotence in males. Diabetic neuropathy affects about half of all diabetics, and there has been no effective treatment for it.

A common complication for diabetics is degeneration of the retina of the eye — diabetic retinopathy. This is the most common cause of blindness in middle-aged people. It happens because diabetes causes swellings in the walls of the arteries feeding the retina and twists in the retinal veins. As a result, there are tiny haemorrhages, and the retinal tissue degenerates and dies.

Evening primrose oil and diabetes

Evening primrose oil is a major breakthrough in the treatment of diabetic neuropathy. Evening primrose oil has been found actually to reverse the nerve damage in diabetics.

It also makes sense to take evening primrose oil to prevent damage to the retina. There have been successful trials in Holland where linoleic acid was used, and it was found that the patients with diabetes who took large amounts of linoleic acid stopped the retinas from degenerating. Evening primrose oil would work in the same way as linoleic acid, but because it is more powerful, you would not need to take as much.

Nerve damage in diabetics

Until now, nerve damage in diabetics could not be treated. It is possible that evening primrose oil will prove to be an effective treatment for diabetic neuropathy, as evening primrose oil has been found actually to reverse the nerve damage in diabetics. This was the conclusion of a study done at the Institute of Neurological Sciences, Glasgow University Medical School.[1]

In this study, a research team in Glasgow specializing in diabetic neuropathy tested evening primrose oil (*Efamol*) on 22 diabetic patients in a double-blind, placebo-controlled trial. Half the patients were given *Efamol* and half were given dummy capsules which looked the same. Neither the doctors nor the patients knew who was receiving which until the end of the six month trial.

The dose used was four 500mg capsules morning and evening.

They were looking at eight different aspects of nerve function. They concentrated on sensations of heat and cold, and on nerve conduction. At the end of the trial all eight aspects of nerve function had improved in the group taking evening primrose oil, whereas all eight aspects had deteriorated in the group taking the dummy capsules.

Eye and heart damage in diabetics

A study at Erasmus University in Rotterdam on linoleic acid [2] makes it likely that evening primrose oil will also slow down the development of damage to the eyes and to the heart in diabetics.

The Dutch researchers conducted an important five-year study. Diabetics were divided into two groups, one taking a diet with four times as much linoleic acid as the other. At the end of the five years, diabetic eye disease was half as common in the high linoleic acid group as in the group taking normal amounts of linoleic acid.

New heart disease was three times as common in the group taking normal amounts of linoleic acid as in the high linoleic acid group. This meant that large amounts of linoleic acid could sharply reduce the risk of long-term eye and heart damage in diabetics.

The Dutch researchers also found that the diabetic patients on the diet high in linoleic acid needed less insulin. In another study, in which patients received a high linoleic acid intake over seven years, diabetic retinopathy was substantially reduced.[3]

Evening primrose oil has not itself been the subject of trials on diabetic retinopathy. But, as GLA — the active ingredient of evening primrose oil — is manufactured from linoleic acid, it would work in the same way. In fact, linoleic acid itself is badly assimilated by diabetics, as will be explained below, whereas there is no problem for diabetics in assimilating evening primrose oil.

How and why evening primrose oil work in diabetics

In diabetics there is something wrong with the way essential fatty acids are metabolized because of the inhibition of a key enzyme — the delta-6-desaturase (D6D).[4] In fact, diabetes is one of the most important known inhibitors of D6D in both animals and in man. Diabetics have low levels of essential fatty acid derivatives in their blood.[4,5,6,7,8,9,10]

The defect in this enzyme puts a spanner in the works at the first step of the conversion process of linoleic acid to gammalinolenic acid. Diabetics have significantly reduced levels of essential fatty acids in their blood.

Evening primrose oil deftly by-passes this enzyme by starting at the next stage of the conversion process.

Obviously, some linoleic acid can get through this blockade, but it does mean taking four times the normal amount to have a good effect. This was proved by the Dutch study on diabetic retinopathy. However, because evening primrose oil does not need to blast its way through any blockade, it is likely to have beneficial effects at normal doses.

Evening primrose oil combined with fish oils for diabetics

The Glasgow study on diabetic neuropathy used evening primrose oil (*Efamol*) on its own.[11,12,13] But investigations are currently being done using evening primrose oil combined with fish oils (as *Efamol Marine*).

The thinking behind this is that because diabetes blocks the conversion of both families of essential fatty acids — linoleic acid and alpha-linolenic acid — it would be sensible to take derivatives of these parent fatty acids, which could by-pass the blocked enzyme.

For the linoleic acid family, this means taking gammalino-lenic acid (i.e. evening primrose oil). And for the alpha-linolenic acid family, it means taking eicosapentaenoic acid (EPA).

As well as being good for diabetic neuropathy and diabetic retinopathy, a combination of evening primrose oil and fish oils is also likely to be good for other complications of diabetes. Such a regime is likely to reduce high cholesterol and triglyceride levels, and to lessen the risk of thrombosis, and also to prevent kidney problems. Further research is planned to test evening primrose oil in these areas.

11 HEART DISEASE, VASCULAR DISORDERS AND HIGH BLOOD PRESSURE

Heart disease and diseases of blood vessels are among the biggest killers in the western world. If ways could be found to reduce the risk of getting them, and to treat them, then countless lives could be saved.

Despite a certain amount of controversy over the underlying risk factors which contribute to coronary deaths, a number are broadly accepted:

- High cholesterol levels in the blood
- Platelets that stick together unduly (platelet aggregation)
- High blood pressure
- Atheroma clogging up blood vessels
- Vascular spasm

Deaths from cardiovascular disease are high in those countries where a lot of saturated fat is eaten.[1,2,3] Studies suggest that diets rich in polyunsaturated fatty acids (PUFAs) do give some protection against these diseases. The death rate in the USA from cardiovascular diseases has fallen by a remarkable 20–25% in the last decade or so, since the population was persuaded to eat less saturated fat and switch to a high PUFA diet.[4,5,6] In countries where this advice has not been followed, such as the UK and Scotland,[7] the death rate from cardiovascular disease remains high.

How evening primrose oil helps to lower cholesterol

Having high cholesterol levels in the blood is one of the broadly accepted risk factors for cardiovascular disease. Total plasma cholesterol remains the best single predictor of coronary heart disease.[1,4,8,9]

Ever since the 1950s it has been known that linoleic acid is able to reduce cholesterol levels.[10,11] However, this would mean taking large quantities of linoleic acid, which would be very high in calories as well as relatively unpalatable.

It is now becoming clear that the power to reduce cholesterol levels is not so much vested in the linoleic acid itself, but in its metabolites dihomo-gammalinolenic acid (DGLA), and also arachidonic acid.[28]

There is a strong association between the incidence of cardiovascular disease and reduced levels of DGLA. A study in Scotland showed that among symptomless middle-aged men, the 10% with the lowest DGLA levels in adipose tissue had an almost 25% chance of developing coronary heart disease over a four year period. Similar results have been obtained in Finland.

In the metabolic pathway of linoleic acid, DGLA is formed from gammalinolenic acid (GLA), the active ingredient of evening primrose oil. Evening primrose oil converts into DGLA very simply. You only need to take six capsules (3g) of evening primrose oil a day to achieve the same effect as taking between 30–100g of linoleic acid! On average, evening primrose oil is over 100 times more potent than linoleic acid in reducing cholesterol.[29,30,31]

DGLA seems to be the key substance in reducing cholesterol. Also, PGE1, which is derived directly from DGLA, has many desirable actions in the sphere of cardiovascular disorders, of which inhibiting cholesterol biosynthesis is only one. PGE1 also lowers blood pressure, and inhibits platelet aggregation and smooth muscle proliferation.[32,33,34]

High cholesterol levels are particularly dangerous because cholesterol not only furs up the blood vessels, but also acts as a blocking agent in the linoleic acid metabolic pathway. Cholesterol interferes with the actions of the delta-6-desaturase enzyme, which helps convert linoleic acid into GLA. This means that people with high cholesterol levels may take a diet high in linoleic acid in the hope of reducing the high cholesterol levels in their blood, but they won't be able to make use of this linoleic acid because of the high cholesterol levels which block its conversion. So they are caught in a catch 22 situation.

However, there are no such problems taking evening primrose oil, as it by-passes this block by starting at the next stage in the metabolic pathway (GLA).

Interestingly, some other blocking agents are also well known risk factors for cardiovascular diseases, such as ageing, diabetes, trans fatty acids (see below), saturated fats and catecholamines released during stress. Perhaps these factors are risk factors for cardiovascular diseases precisely because they block the pathway of linoleic acid. Once again, evening primrose oil circumvents all these blocking agents by starting at the next step, GLA.

Trans fatty acids

Trans fatty acids have a harmful effect on the cardiovascular system. In order for fatty acids to be biologically useful, they have to be in what is called the 'cis' form. Once they are processed, they may lose their 'cis' form and become 'trans'. Once they are trans they behave like saturated fats. Not only that, but they actually compete with cis-linoleic acid and so inhibit its metabolism.

People in industrialized countries eat on average 6–12g of trans fatty acids a day, and these fatty acids are found in substantial amounts in human tissues.[31,35,36,37]

Apart from acting as a blocking agent in the metabolic pathway of linoleic acid, trans fatty acids are also known to raise cholesterol levels. So taking large quantities of trans fatty acids over a long period of time is likely to have harmful effects on anyone who is a candidate for cardiovascular disease.

Evening primrose oil can by-pass the block created by trans fatty acids and can also help to lower cholesterol. However, it is far better to take evening primrose oil as part of a diet low in saturated fat and trans fatty acids.

Studies on evening primrose oil and cholesterol

Evening primrose oil has an interesting effect on cholesterol levels. It will only bring down cholesterol levels if they are high, but it will have no effect on cholesterol levels if they are normal or low. As the starting cholesterol levels rise, so the relative potency of the GLA in evening primrose oil increases.

Evening primrose oil is clearly an effective cholesterol-lowering agent in those people with plasma cholesterol values above 5 mmol/l, in other words, in all but the lowest 20% of cholesterol levels.[29]

Like other PUFAs, evening primrose oil either has no effect on HDL (high density lipoprotein) cholesterol or actually increases it. The cholesterol-lowering action of evening primrose oil is entirely because it is able to lower LDL (low density lipoprotein) cholesterol. It is the LDL cholesterol which is harmful and which needs to be lowered in order to reduce the risk of a heart attack.[38] HDL cholesterol is desirable because it actually helps to transport cholesterol away from places where it may be harmful. Evening primrose oil has the very beneficial effect of raising the HDL/LDL ratio.

Evening primrose oil and platelets

A risk factor for cardiovascular disorders occurs when the platelets in the blood aggregate abnormally — they bunch up and stick together. Evening primrose oil is very effective in stopping this process.[19,20,21]

The clotting agents in the blood are called platelets. And they are called platelets simply because these tiny particles look a bit like plates. When you cut yourself the blood runs out of the open wound for a short while, but then a clot forms to stem the flow. This clot is formed because the platelets stick together, which is an entirely desirable thing for them to do when you suffer a skin wound.

But the problem starts when the platelets start to bunch up and stick together when you haven't suffered any skin wound. In healthy people, the blood only feels sticky when it begins to clot after a skin wound. While it is coursing round inside you the blood would feel slippery if you could touch it. But the blood inside you should not be sticky. In recent heart attack victims their blood is about $4\frac{1}{2}$ times stickier than in normal people. If you looked at their blood under a microscope you would see platelets sticking to each other and to artery walls.

When platelets stick to cholesterol deposits this quickly leads to a clot, which can block the flow of blood. When a blood clot forms in an artery or a vein, it's called a thrombosis. This blocks the circulation in the area. A clot in a coronary artery is a coronary thrombosis. In the brain it's a stroke.

Normally, artery walls make prostacyclin which prevents platelets from sticking to each other or to artery walls. But arteries which are damaged by fat and cholesterol deposits, high blood pressure or injury, don't make enough prostacyclin.

Evening primrose oil helps because the GLA in evening primrose oil easily converts to DGLA, which is known to be able to reduce platelet aggregation. If the platelets can stop clumping together, the risk of thrombosis is reduced. Also, evening primrose oil converts to prostaglandin E1, and PGE1 is one of the most potent known inhibitors of platelet aggregation.[27]

Evening primrose oil and blood pressure

This is the pressure at which the heart pumps blood into the major arteries. If the blood pressure goes too high, it overtaxes the heart and blood vessels. People with high blood pressure run a greater risk of experiencing arteriosclerosis, heart failure, stroke, and kidney disease.

Tests on both animals and humans have shown that essential fatty acids reduce arterial pressure.[12,13,14,15,16,17,18] Evening primrose oil has been shown to bring down the blood pressure in animals with high blood pressure. In preliminary studies on humans with high blood pressure, evening primrose oil (given as *Efamol*) was considered more effective in lowering blood pressure than much higher doses of other polyunsaturates.

Evening primrose oil and vascular obstruction

Diets rich in polyunsaturated fatty acids may not only arrest the progression of atheroma, but may actually reverse it, allowing the obstruction to be cleared.[22,23] Taking a supplement such as evening primrose oil would be worth while even for those people whose cardiovascular system is already damaged. In a group of people with intermittent claudication due to vascular obstruction, evening primrose oil was found to improve their exercise tolerance.[24]

Evening primrose oil helps in two ways. Firstly, because the GLA in evening primrose oil converts easily to DGLA, and it is likely that many of the beneficial effects of essential fatty acids in cardiovascular disease may relate to an accumulation of DGLA. Secondly, evening primrose oil converts to PGE1 which is a potent vasodilator — it widens the peripheral blood vessels. PGE1 has produced dramatic improvements in patients with vascular spasm due to Raynaud's syndrome (see Chapter 15), and also to relieve angina.[25,26,27]

Drugs and cardiovascular disorders

Several drugs are on the market to treat conditions mentioned in this chapter. However, evening primrose oil works physiologically rather than as a drug, and with none of the side effects of drugs. The fact that it works physiologically rather than pharmacologically is proved by the fact that evening primrose oil does not lower cholesterol in those people with normal levels of cholesterol — it acts through natural processes to regulate cholesterol metabolism.

Fish oils

Fish oils have had a remarkable success in trials with patients suffering from angina, hyperlipidaemia, and those having suffered a heart attack. A number of studies have shown that when you add fish oils to your diet, cholesterol levels are lowered and there is less clumping together of platelets.

So it is a good idea to take a brand of evening primrose oil which already contains fish oils, or to take a supplement of fish oils in addition to evening primrose oil.

12 MULTIPLE SCLEROSIS

The management of multiple sclerosis (MS) involves changing one's diet and lifestyle and also taking specific supplements. This regime can help someone with MS improve, rather than get worse. Evening primrose oil is a very important nutritional supplement in the management of MS.

In particular, the kind of fat which someone eats seems to have a strong bearing on MS.

The geographical distribution of MS

One of the most marked features of MS is its geographical distribution. MS is a disease of temperate zones, and is virtually non-existent in the tropics. One of the key differences between areas of high and low incidence of MS seems to be the food that people eat.

In those places where MS is commonest, people eat a lot of dairy produce. In those places where MS occurrence is lowest, people eat more fish and vegetable oils. The difference between an area of high MS and low MS can be as little as a few miles. So some of the starkest contrasts in MS distribution are in Norway where MS is high in inland farming areas where dairy farming is practised and low in costal areas where people eat a lot of fish. Similarly in some Scottish islands, the rates of MS can fluctuate from very high to very low according to the main diet of the local people — high in areas of dairy farming and low in fishing areas.

One of the first doctors to look at the world map of MS was Professor Roy Swank, now based in Portland, Oregon, USA. He first developed his famous Swank Low Fat Diet[6] in 1948. Swank noticed several important clues. First, the amount of saturated fat in the typical American diet was rising dramatically. And as the consumption of saturated fat increased, so did the incidence of certain diseases — particularly MS, heart disease and stroke.

There were further clues for any medical detectives on the look-out during World War II. It was noticed that young American soldiers who had died of heart attacks during training and battle showed a greater degree of hardening of the arteries than their Oriental counterparts who ate mostly vegetables and rice.

In occupied Norway, fat consumption fell by 50% during food shortages. At the same time, there were significant reductions in death rates from heart attacks, and the rate of MS dropped too.

In the UK today, 40% of our diet is saturated fat. Around the world, tipping the balance of saturated/unsaturated fat in favour of saturated fats has coincided with an increase not just in MS, but also in cardiovascular (heart and stroke) diseases.

The amount of sugar we eat has increased enormously in the same period of time. Some nutritionists believe that humans were not designed to thrive on a high saturated fat plus high sugar diet. The rise in chronic disease coincides with these radical changes in diet in the western world.

Fats and the composition of the brain, the nervous system, and cell membranes

The importance of unsaturated fats becomes clear when you look at the composition of the brain, the nervous system, and cell membranes. Roughly 60% of the brain is made up of structural fats. These special fats, called phosphoglycerides, are the building blocks of the central nervous system and are very rich in essential fatty acids. The essential fatty acids cannot be made within the body but must be obtained from the diet. That is why they are called essential.

Cell membranes also need these essential fatty acids (as well as proteins) to be built properly. Cell membranes must have fluidity and flexibility to be in good shape. They get this fluidity and flexibility from essential fatty acids.

So overall, the body has a great need for essential fatty acids. That's why a diet high in polyunsaturated fats and low in saturated fats is so important for MS. Evening primrose oil is a very rich source of polyunsaturated fats.

Why polyunsaturated fatty acids are important in MS

- Tests have shown that the levels of linoleic acid in the plasma tend to be low in some — but not all — MS patients, and they go even lower during an exacerbation.[12] (Low linoleic acid levels can also be found in several other diseases.)
- They are needed for the growth and repair of nervous tissue.
- They are needed for the maintenance and structure of nervous tissue. This is particularly important in MS, where the nervous system is under attack. If the body lacks these nutrients, any repair of damaged tissue is made more difficult.
- People with MS show an unusual pattern of fatty acids in their blood. With a diet rich in PUFAs, this can return to normal within a few months.
- Some research has shown that the white matter in the brains of people with MS is low in PUFAs.
- Perhaps people with MS have an inborn inability to handle PUFAs correctly.
- In people with MS, the myelin sheath, the red and white blood cells, the platelets, and the blood plasma are also deficient in PUFAs, particularly linoleic acid.
- The activity of lymphocytes (white blood cells) may be dependent on the state of the cell membrane. They will behave differently according to whether a cell membrane is fluid (plenty of PUFAs) or rigid (not enough PUFAs). This influences the ability of certain lymphocytes to react immunologically.

Trials involving polyunsaturated fats and MS

Thanks to the work of scientists like Professor R.H.S. Thompson, Professor Roy Swank and Professor Hugh Sinclair, who observed the link between saturated fat and multiple sclerosis, doctors working in the field of multiple sclerosis thought it was worth investigating polyunsaturated fats further.

The first big trial involving linoleic acid and MS was conducted in 1973 by Dr J.H.D. Millar of Belfast and Dr K.J. Zilkha of the National Hospital in London, and others. They found that when linoleic acid, in the form of sunflower seed oil, was given to patients with MS, it reduced the frequency and severity of relapses.

After this trial, sunflower seed oil in various forms became all the rage with MS patients. They drank it neat, they took it in emulsions, they mixed it with orange juice. Many of them didn't like it.

At this time, evening primrose oil capsules were being manufactured by one company only, Bio-Oil Research Ltd, of Cheshire. It was Bio-Oil's director, John Williams, who was the first to see the potential of evening primrose oil, originally for heart disease. But when the results of the sunflower seed oil trial were published in the *British Medical Journal* in 1973,[1] John Williams had a brainwave. If sunflower seed oil helped a little, then surely evening primrose oil, being that much more biologically active, might help even more.

At around the same time, Professor E.J. Field was doing some very important research work on essential fatty acids and MS. He started this research while Director of the Medical Research Council's Demyelinating Disease Unit in Newcastle, and later carried on with the research at Newcastle University. Professor Field tested evening primrose oil on the red blood cells of people with MS.[2] The results of these blood tests proved that the gammalinolenic acid (GLA) in evening primrose oil was much better than linoleic acid in correcting the defects found in the blood of MS patients.

Note: I was one of the MS patients whose blood was tested by Professor Field. After almost a year of taking evening primrose oil capsules, my EFA blood abnormalities were corrected. I was not on any particular diet at the time.

Trials on evening primrose oil and MS

During the 1970s many patients with MS began to take evening primrose oil, without the oil ever having been tested in a trial. Then in 1978 a trial took place in Newcastle, conducted by Professor David Bates and others.[3]

The researchers divided 116 people with MS into four groups. One group was given evening primrose oil (*Naudicelle*), six capsules a day; one group was given olive oil in capsules; one group was given 'Flora' to eat as a spread; and one group was given another spread. No one knew what he or she was taking.

At the end of the two years, there was no significant difference between any of the groups, as measured by the Kurtzke disability scale.

Of all the groups, those who did best were the ones who took the sunflower seed oil spread 'Flora'. The duration and severity of their attacks were less severe. In this group, the amount of linoleate in their blood went up from 28% before the trial started to 39% at the end of the trial.

Professor Bates came to the conclusion that the amount of polyunsaturates taken has to be enough to affect plasma levels. Only when this level has been achieved does the PUFA have an effect on the severity and duration of relapses.

At the time, the results of this trial were taken to prove that evening primrose oil does not work for MS. But this is not a correct or fair assessment at all, and in fact the results of this Newcastle trial were later re-analysed by a Canadian doctor by the name of Robert Dworkin. Some years later after the Newcastle study, Dr Dworkin looked closely at its results, but he also pooled these results together with results from two other trials, one jointly in London and Belfast (Millar and others) and one in Ontario.[4]

What Dr Dworkin found was extremely important: *patients who had very low levels of disability at the start of the trial, who took polyunsaturates, did not get worse in a two-year period.*

This was indeed a crucial discovery — the length of time that a patient had had MS made a difference to the outcome of the trials. The newly-diagnosed, who were 0–2 on the Kurtzke disability scale at the start of the trials, were the ones who showed little change or deterioration by the end of the trial. This applied only to the group which had been treated with PUFAs.

The conclusion from this is that *treatment with PUFAs helps to stabilize MS in the recently diagnosed who have no real disability.*

So in fact the Newcastle study — far from showing the ineffectiveness of evening primrose oil and other polyunsaturates for MS — does the opposite. But it does show that someone with MS does need to take a certain amount of linoleate for it to be effective.

The trial results do show that six capsules of evening primrose oil on their own, without any additional intake of linoleic acid, is not enough to affect plasma levels of linoleate. The answer, surely, is to take eight to 12 capsules of evening primrose oil a day, plus use a sunflower seed oil spread and a cooking oil high in linoleic acid.

Some people have also criticized this particular trial on other scores. Firstly, there was no advice given about cutting down on saturated fats in the diet. (Saturated fats are thought to compete with polyunsaturated fats.) Secondly, the *Naudicelle* capsules used at that time had orange and black coloured shells which used the dye tartrazine. It is known that tartrazine interferes with fatty acid metabolism. Since then, evening primrose oil capsules have been produced in clear gelatin shells, with none of the same problems.

It is a pity that no one has conducted another trial with evening primrose oil taking all these factors into consideration. Since there has been no scientific evidence in favour of evening primrose oil as a therapy for MS, it is not prescribable on the NHS and has to be bought from chemists or health food shops, or by mail order from the manufacturers. Many people with MS find evening primrose oil too expensive to buy so don't take it at all. (Discounts are possible, see pages 122-123.)

Why evening primrose oil is better than linoleic acid

The results of the 1978 Newcastle study and the 1973 London and Belfast study might lead one to believe that sunflower seed oil, sunflower seed oil spreads and similar sources of linoleic acid were the most effective PUFAs for MS.

It is indeed a very good idea to use such seed oil products in one's daily diet. However, there are still very good reasons for supplementing the diet with evening primrose oil.

The most important reason is that, no matter how much linoleic acid you consume, there could be some problems metabolizing it. Unless the linoleic acid is converted inside the body to longer-chain fatty acids and prostaglandins, it is of little use in MS.

Many things can get in the way of converting all the linoleic acid you eat. These are the *blocking agents*. The most common ones are:

- Foods rich in saturated fat
- Foods rich in cholesterol
- Foods rich in trans fatty acids
- Stress, when the body releases catecholamines
- A diet too high in simple sugars
- Low levels of zinc, magnesium, Vitamin C, Vitamin B6, nicotinamide (B3), and biotin
- Alcohol in moderate to large amounts
- Viral infections
- Atopic conditions

Figure 8. The ways in which PUFAs may be working in the treatment of MS.

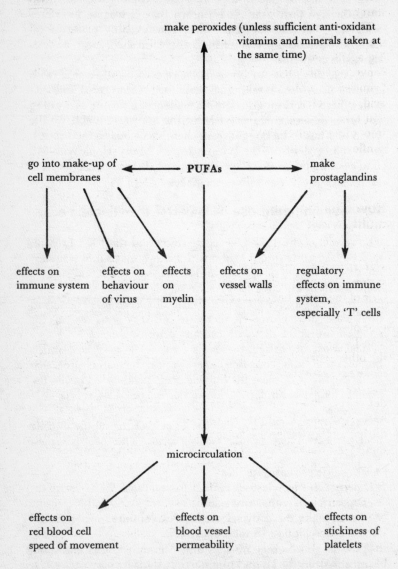

Note: I am grateful to Professor Bates, who conducted the Newcastle trial in 1978, for the above figure. Ironically, he is now convinced of the benefits of polyunsaturates for MS.

There are very few people with MS who do not have several of these 'blocking agents' operating in their own lives, before they start any kind of self-help therapy.

The virtue of evening primrose oil is that it begins later on in the metabolic pathway, completely avoiding most of the blocking agents which for the most part inhibit the first step — the conversion of linoleic acid to gammalinolenic acid. Evening primrose oil, unlike sunflower seed oil, is rich in gammalinolenic acid, thus starting at step 2 in the metabolic pathway. So when you take evening primrose oil, you can be sure it will convert into the longer-chain fatty acids, and prostaglandins. With sunflower seed oil, you cannot be sure of this at all.

How evening primrose oil might be working in multiple sclerosis (see figure 9)

1 It stimulates the T-lymphocytes (i.e. it boosts the immune system). Evening primrose oil converts into Prostaglandin E1. One of the many beneficial roles of PGE1 is to activate defective T-lymphocytes.[7] It stimulates the normal functioning of this type of white blood cell, which is thought to be defective in MS. The T-suppressor lymphocytes are white blood cells which keep the other parts of the immune system under control and which make sure that the body's defences attack foreign materials and not the body's own tissues. When T-suppressor cells are defective, auto-immune damage frequently occurs. Research has shown that T-suppressor cells are very low in MS patients during a relapse, and PGE1 may help prevent this. It is known that PGE1 has the effect of dampening down the B-lymphocytes which are capable of attacking the central nervous system.

2 It strengthens blood vessel walls. Prostaglandin E1 is known to strengthen blood vessel walls. This is particularly important in MS because there is growing evidence that in the microcirculation of people with MS the blood vessel walls are breached so that blood — which is toxic to nerve tissue — seeps into the brain. It crosses the blood/brain barrier. If the blood vessel walls are strengthened, they are better able to withstand things like platelets and cholesterol clumping together and sticking to the walls.

Figure 9. The bumpy metabolic road of cis-linoleic acid in MS.

STEP 1　**CIS-LINOLEIC ACID**

↓

enzyme delta-6-desaturase needed to get to step 2

↓

helped by zinc, magnesium, vitamin B6, biotin

↓

BLOCKED BY : saturated fats
　　　　　　　cholesterol
　　　　　　　trans fatty acids
　　　　　　　stress
　　　　　　　too much sugar
　　　　　　　low levels of zinc, magnesium, B6, biotin,
　　　　　　　　　Vitamin C, nicotinamide
　　　　　　　viral infections
　　　　　　　too much alcohol
　　　　　　　atopic conditions

↓

STEP 2　**GAMMALINOLENIC ACID** (EVENING PRIMROSE OIL
　　　　　　　　　　　　　　　　STARTS HERE)

↓

STEP 3　**DIHOMO-GAMMALINOLENIC ACID**

↓

helped by vitamin C, nicotinamide (B3)

↓

STEP 4　**PROSTAGLANDIN E1**

Some people believe that partly-digested food is able to get through the intestine walls in people with MS, which may be a factor in food allergies. PGE1 may help here too.

3 It stops the platelets clumping together. In MS there is evidence that the platelets, the small plate-like particles in the blood which help the blood clot, clump together in an abnormal way. PGE1 regulates the platelets and stops them bunching up together, sticking to each other, and to blood vessel walls.

4 It makes faulty red blood cells return to normal. In MS red blood cells are not only very low in essential fatty acids, they are also much bigger than they ought to be, and have a poor ability to regulate the passage of fluids through cell membranes. Evening primrose oil can correct this defect within a matter of months.

Evening primrose oil has also been shown to correct the defect in the mobility of red blood cells. Electrophoretic mobility tests have shown that the red blood cells of people with MS move more slowly than those of healthy people.[8] After several months of evening primrose oil supplements, these red cells have been shown to behave normally.

In a follow-up study of MS patients on long-term treatment with evening primrose oil (*Naudicelle*), who were also following a diet low in saturated fat, it was found that the mobility of red cells returned to normal.[9] The most responsive cases were those who had experienced frequent relapses.

5 It possibly acts as an anti-viral agent. When human cells become transformed by viruses, they usually lose the ability to convert linoleic acid to gammalinolenic acid, and so can't make PGE1.[10] This may make the transformed cells resistant to attack by the body's natural defences, the immune system. Evening primrose oil gets over this problem by its ability to convert easily to PGE1. PGE1 may restore the cells' normal susceptibility to the body's immune system.

6 It affects the nervous system. Evening primrose oil provides the kind of structural fats which go to make up parts of the central nervous system. Essential fatty acids are needed for myelin, the nerve sheath which breaks down in MS, but which can be regenerated.

Evening primrose oil affects not only cell membranes but also nerve conduction and the action of nerves, via PGE1. This can produce profound changes in the workings of both the central nervous system and the peripheral nervous system. PGE1 has strong regulating effects on the release of neurotransmitters at nerve endings and also on the post-synaptic actions of the released transmitters.

7 It maintains a healthy balance between the 1 and 2 series prostaglandins. If the body is very low in essential fatty acids, there is a sharp rise in the 2 series PGs, which are made from arachidonic acid. A high level of the 2 series PGs is a feature of various inflammatory disorders, such as rheumatoid arthritis and possibly MS. It has been shown that the cerebrospinal fluid from MS patients contains high levels of PGF2alpha.

Once you increase the amount of essential fatty acids in the diet, PGE1 is back on the scene. Enough PGE1 means that there is a healthy balance in the amount of 1 and 2 series PGs being produced.

Another thing that evening primrose oil does is to make it more likely that PGE1 will be produced, as against PG2. In the metabolic pathway of linoleic acid, it encourages the route towards PGE1 at the junction where the road forks after dihomo-gammalinolenic acid.

Nutritional supplements to be taken with evening primrose oil

Evening primrose oil should not be taken just on its own, not only in cases of MS, but in any condition.

Ideally, it should be taken with certain co-factors, which help in the metabolism of essential fatty acids. These are:

- Vitamin C
- Vitamin B6
- Nicotinamide (Vitamin B3 or niacin)
- Zinc
- Magnesium

In addition, evening primrose oil should always be taken with Vitamin E. Vitamin E acts as an anti-oxidant, to prevent harmful peroxides.

Many evening primrose oil capsules contain Vitamin E for this very reason, but some of the cheaper brands do not. So if

Figure 10. How evening primrose oil favours the PGE1 route.

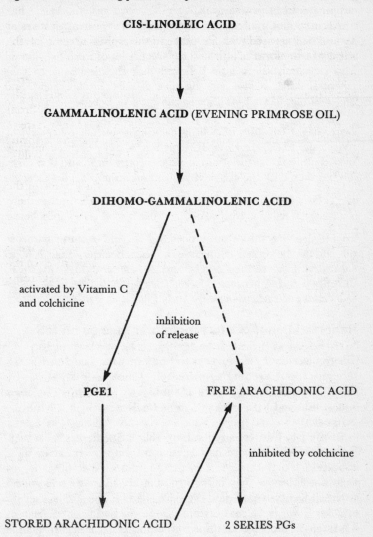

you buy a brand which is without Vitamin E, be sure to take a supplement of this vitamin.

An intake of linoleic acid in the form of evening primrose oil should be balanced with an intake of the alpha-linolenic family, ideally in the form of a fish oil capsule. A lot of fresh fish should also be eaten. It is now believed that the best way to take essential fatty acids is a balance of the linoleic acid and alpha-linolenic acid families in the ratio of between 3 and 6 to 1. [3-6]

Dose. Eight capsules of 500mg evening primrose oil a day, or more, up to 12 or even 16 capsules a day, divided equally between breakfast-time, lunch-time, supper-time and bed-time. To be taken with the above vitamins and minerals, plus fish oils.

Diet

Diet is also very important indeed in MS, and evening primrose oil should be taken alongside a very healthy diet, low in saturated fats, refined foods and sugar, and high in fresh, unrefined and unprocessed foods. (See *Multiple Sclerosis — A Self-help Guide to Its Management* by Judy Graham, Thorsons, 1987).

Reported benefits of evening primrose oil on MS

There have been two surveys to find out what people with MS thought about evening primrose oil. One was carried out by Bio-Oil Research Ltd.,[11] who manufacture *Naudicelle*. The other was conducted by ARMS (Action for Research into Multiple Sclerosis).

In the Bio-Oil Research survey 480 MS sufferers took part. Sixty-five % felt there had been some improvement in their condition. Forty-three % said there had been a stabilization of their condition — they had got no better, and they had got no worse. Twenty-two % said there had been fewer and less severe attacks. Twenty % said certain symptoms had been alleviated. Thirteen % reported an improvement in general health. People in the 'some improvement' category mentioned the following benefits:

• Increased mobility
• Increased walking ability
• Reduced spasm or tremor

- Improved bladder function
- Improved eyesight
- Improved condition of hair and skin
- Relief of constipation
- Improvement in wound healing
- Regaining correct weight
- Heavy periods returned to normal

Note: The improved group contained a significantly higher proportion of MS patients who had been diagnosed within the preceding four years.

In the ARMS survey, out of 177 completed questionnaires, 127 said they had improved, 33 reported no change, and 17 felt that their symptoms became worse.

Even though these surveys have no scientific standing, and all the answers are based only on the subjective opinion of the MS sufferer who filled in the questionnaire, the results are nevertheless extremely encouraging.

ARMS members were also asked how long they had been taking evening primrose oil. The answers showed that improvements increased when they had been taking the capsules for more than four months. Beneficial effects appeared as follows:

Under 4 months	35%
4 months to 1 year	73%
1 to 2 years	73%
2 to 3 years	82%

(At the time of the survey, very few members had been on *Naudicelle* for longer than three years.) Of the people who returned the completed questionnaires, 141 were also on some kind of diet. The results showed that the people who were exercising some control over their diet (i.e. less saturated fat etc.) had better results with the evening primrose oil.

13 ALCOHOLISM

Evening primrose oil has proved to be very useful in treating certain aspects of alcoholism, particularly withdrawal symptoms, liver damage, and hangovers. It may also help reverse some of the brain damage caused by alcoholism.

Alcoholism and Prostaglandin E1

The reason why evening primrose oil works in alcoholism is largely because it converts to Prostaglandin E1 (PGE1). This prostaglandin plays a key role in alcoholism.

Alcohol has a paradoxical effect on PGE1. When you drink small amounts of alcohol, it stimulates the production of PGE1. But when the level of drinking is stepped up, alcohol robs the body of PGE1.

In fact, small amounts of alcohol may be positively good for you. Recent research in North America and Western Europe has consistently reached this conclusion. The risk of heart disease is reduced in people who have a modest intake of alcohol — perhaps a glass or two of wine, or a beer, or one or two drinks of whisky a day. People who drink at this level tend to live longer than teetotallers, and much longer than people who drink more alcohol.

Many people drink socially because it makes them feel good. This may well be because, without knowing it, they are raising their levels of PGE1.[1] PGE1 can bring about euphoric states: doctors have noticed almost euphoric reactions in patients being given intra-venous infusion of PGE1. And studies on manic people, who feel abnormally euphoric, show that they make more PGE1 than normal.[2] However, this mood-elevating effect of PGE1 from drinking alcohol only happens when someone drinks small amounts.

The story changes quite dramatically once the intake of alcohol increases to moderate or large amounts. At this level, alcohol has two bad effects on the body's stores of essential fatty acids, and therefore on the metabolism of prostaglandin E1.

First, alcohol enhances the conversion of dihomo-gamma-linolenic acid to PGE1. This may sound like a good thing, but in fact it has disastrous consequences. Because alcohol stimulates the synthesis of PGE1,[2,9] it draws on the body's stores of dihomo-gammalinolenic acid (DGLA), and these stores of DGLA can run out quite quickly. When the action of the alcohol wears off, the stores of DGLA may be so reduced that the levels of PGE1 fall below normal.

Second, alcohol in moderate to large amounts is one of the main blocking agents in the metabolic pathway of essential fatty acids. Alcohol blocks both the delta-6-desaturase enzyme,[10] and also the delta-5-desaturase enzyme.[3] The delta-6-desaturase enzyme converts linoleic acid to GLA, and the delta-5-desaturase enzyme converts DGLA to arachidonic acid. These two enzymes are also involved in the metabolism of the alpha-linolenic acid family, and alcohol blocks the metabolic pathway of this family of essential fatty acids as well. Incidentally, alcohol has no effect on these enzymes if it is only drunk in small amounts.

The consequences of these two things are serious. As alcohol blocks the conversion of linoleic acid to GLA, this also has an effect on the stores of DGLA. It means that no matter how much linoleic acid you eat in your diet, the depleted stores of DGLA cannot be replenished. Consequently, drinking moderate to large amounts of alcohol over a long period of time will lead to a serious deficiency of essential fatty acids.[11]

Many of the long-term effects of alcohol may relate to these very low levels of essential fatty acids. The dramatic fall in PGE1 which happens in alcoholism may account for the hangovers, withdrawal symptoms and depression that so often go with heavy drinking.

The low levels of PGE1 in alcoholics may have other serious consequences too. This includes the risk of heart attacks and strokes, high blood pressure, a reduced ability to cope with infections, brain and nerve deterioration, and liver damage.

An alcoholic gets caught in a vicious spiral. He or she may have started drinking because it gave his or her mood a lift. But the more he or she drinks, the more he or she needs to drink to get the same 'high'. The explanation for this is that as you drink more and more, your supplies of dihomo-gammalinolenic acid get lower and lower, so PGE1 simply cannot be made.

Evening primrose oil and the problems of alcoholism

Evening primrose oil works in alcoholism because it is rich in gammalinolenic acid. This means it can avoid the enzyme block which prevents linoleic acid from converting to GLA. It increases the body's supplies of essential fatty acids, and its store of DGLA, and means that PGE1 levels can be raised.

These properties make evening primrose oil useful in a number of conditions associated with alcoholism.

Withdrawal symptoms. Evening primrose oil can alleviate some of the symptoms usually associated with withdrawal from alcohol. In a series of studies conducted by Dr Iain Glen of the Highland Psychiatric Research Group at Craig Dunain Hospital in Inverness, Scotland, patients treated with *Efamol* while withdrawing from alcohol did much better than the patients on a placebo.[4,5,6]

Efamol was found to reduce the amount of tranquillizers needed by alcoholics in the throes of withdrawal. There was also a marked difference in the essential fatty acid content of the plasma and red blood cells after 24 weeks of treatment on *Efamol*, compared with the group given a placebo. *Efamol* also lowered the incidence of hallucinations during the withdrawal phase.

This study on human alcoholics confirms earlier work done on mice by Dr John Rotrosen and Dr David Sagarnick at New York University,[2] who got mice addicted to alcohol by giving them an alcohol-rich diet. They then took away the alcohol abruptly and over the next few hours there was a dramatic withdrawal syndrome, similar to what happens with human alcoholics. The doctors then injected either PGE1 or *Efamol* into the animals. This dramatically alleviated the withdrawal problems of the addicted mice. Tremor, irritability, over-excitability and convulsions were all reduced by about 50%.

Liver and other tissue damage. A common complication of alcoholism is fatty degeneration of the liver. Another study done by Dr Iain Glen in Inverness, Scotland,[4,8] showed that *Efamol* can go a long way towards correcting liver damage due to alcohol. The Alcoholic Clinic at Craig Dunain Hospital conducted a double-blind trial with about 100 patients. No one knew who was taking the capsules of evening primrose oil, and who was taking the identical capsules containing liquid paraffin.

The group taking the evening primrose oil (*Efamol 500*) did much better than the others. The results showed that evening primrose oil can improve liver function and its biochemistry can return to normal much more quickly, compared with a group of alcoholics who were given the placebo.

Hangovers. Evening primrose oil is highly effective in preventing hangovers. Doctors researching this treatment have tried this for themselves, and found that four to six capsules straight after drinking and before going to bed greatly reduce the symptoms of a hangover.

Damage to the brain and the central nervous system. Alcoholism can lead to chronic changes in the brain, and can affect the central nervous system. These changes are associated with the gross lack of essential fatty acids in alcoholics. Alcoholics do worse than other people in particular exercises which test their brain, such as problem-solving, learning new things, memory, perceptual-motor speed, and thinking in an abstract way. Their performance in these tests varies from mild to moderately severe.[4]

It makes sense to treat these changes with essential fatty acids for several reasons. First, essential fatty acids are major structural components of the brain. Second, essential fatty acids have major effects on cell membranes, and thereby can influence nerve conduction, transmitter release, and transmitter action.[7]

Third, prostaglandins, which are derived from EFAs, have profound effects on behaviour. They are also able to modify conduction and transmitter function.[7] This is the rationale for using evening primrose oil, as it is a rich source of essential fatty acids, and it converts to PGE1.

A large proportion of chronic alcoholics and even heavy social drinkers have been shown to suffer brain shrinkage. There can be a significant loss of brain tissue in chronic alcoholics. Even social drinkers may not be free of this minimal brain damage, and can lose their ability to think clearly.

Alcohol also effectively accelerates the ageing process. Young alcoholics, in their thirties to late forties, perform in a similar way to older non-alcoholics who are in their fifties.

It is possible to reverse these changes caused by alcoholism if someone gives up drinking, but it does take time. But exactly to what extent these problems are reversible largely depends on how old the person is and how long he or she has been drinking. Even after one year of abstinence, some alcoholics may still have problems in some areas, such as less practical problem-solving and non-verbal abstracting tasks.[4]

There is evidence to show that alcoholics with these brain and behavioural problems are very low in essential fatty acids. The hypothesis is that taking supplements of essential fatty acids once someone has given up alcohol will help their brain powers and their behaviour return to normal.

Tolerance and addiction

People who crave alcohol need to drink more and more in order for it to have the same effect on their mood, because they have become tolerant to alcohol, and this increased tolerance is a major factor in addiction.

In animal studies, tolerance could be prevented by giving evening primrose oil (*Efamol*) daily with alcohol. If evening primrose oil can prevent the development of tolerance, then maybe it can also help addiction.

Foetal alcohol syndrome

A wide range of foetal abnormalities can be produced by alcohol. These include being underweight and underlength at birth, with slow growth and failure to thrive even with special care. In humans, the babies tend to have unusually small heads, with defective development of some features, such as eyes set very wide apart. There can also be mental backwardness, behavioural problems, and extreme nervousness.

There is some evidence to suggest that a block in the conversion of linoleic acid to gammalinolenic acid because of alcohol maybe the cause of many of the alcohol-induced abnormalities. When evening primrose oil (*Efamol*) was given to laboratory animals who had been given alcohol during pregnancy, most of the abnormalities were not seen.

CAUTION: Pregnant women, of course, should *not* drink.

14 SCHIZOPHRENIA

Evening primrose oil has a significant effect on certain aspects of schizophrenia, particularly on the 'flat' emotions and social withdrawal common in schizophrenics. It also improves overall psychiatric status and memory, as well as helping reduce the abnormal, involuntary movements (tardive dyskinesia) in schizophrenic patients who are on neuroleptic drugs.

Schizophrenia and essential fatty acids

Schizophrenics have abnormalities in essential fatty acids. They are low in the 1 series prostaglandins, and high in the 2 series prostaglandins. No one knows exactly what causes these abnormalities but they seem to be related partly to abnormalities in the essential fatty acid precursors of prostaglandins. In schizophrenics, the levels of linoleic acid, dihomo-gammalinolenic acid (DGLA) and arachidonic acid in red blood cell membranes are below normal.

There also seems to be an abnormality in both the delta-6 and the delta-5-desaturase enzyme, which would inhibit the conversion of linoleic acid to GLA, and from DGLA to prostaglandin E1. The arachidonic acid may be being used up at an excessive rate to form 2 series prostaglandins.

Since schizophrenics are low in essential fatty acids and PGE1, it makes sense to try evening primrose oil as part of any treatment, as it is a rich source of EFAs, and as it converts to PGE1.

Evening primrose oil in the treatment of schizophrenia

There was recently a double-blind trial on *Efamol* on abnormal movements,[1] psychiatric status and memory on patients with

tardive dyskinesia, which showed that evening primrose oil
produced significant results in certain aspects of schizophrenia.
Memory, and total psychopathology scores improved significantly, and there was a marginal improvement on involuntary
abnormal movements.

Tardive dyskinesia (involuntary abnormal movements). Up to
60% of patients on neuroleptic drugs suffer from motor disorders such as tardive dyskinesia as a side-effect. These involuntary abnormal movements are severely disabling and intract-
able.

In a preliminary study by Dr Vaddadi, [2,3] it was noticed that
there was a sudden reduction in these involuntary movements in
some of his schizophrenic patients while they were being treated
with evening primrose oil and penicillin. These early findings
have been backed up by further studies.[4,5,6,7]

The success of Dr Vaddadi's work led to a full-scale double-
blind trial involving several centres [1]: the Department of
Psychiatry at Crawley Hospital, West Sussex, St George's
Hospital Medical School, London, the Department of Psychology, Middle East Technical University, Ankara, Turkey, and
the Efamol Research Institute in Kentville, Nova Scotia,
Canada. As well as researching the effects of evening primrose
oil on tardive dyskinesia, they also looked at the oil's effects on
psychiatric status, and on memory.

The trial involved 48 psychiatric patients with established
movement disorders who had been on neuroleptic drugs over a
long period of time. Thirty-nine of these patients were schizo-
phrenic, four manic-depressive, and three had personality
disorders. All of them had involuntary abnormal movements of
at least mild severity. Patients were kept on their existing
medication, and were matched against controls.

The patients were given either *Efamol*, or a placebo, for 16
weeks. The dose was 12 capsules of *Efamol* (or a placebo) divided
up over the day. At the end of the 16 week period, there was a
cross-over, so that the patients who had been on the *Efamol*
received the placebo instead, and vice versa.

Assessments were made at the beginning of the trial and then
at two-weekly intervals throughout the trial. Blood samples were
also collected and analysed.

The results of these blood tests showed that all psychiatric
patients had below-normal levels of essential fatty acids. Patients

with severe tardive dyskinesia had the lowest levels — significantly lower than both the patients without any movement disorder, and also those with only mild movement disorder.

This shows that there is an association between low levels of essential fatty acids, and the presence and severity of tardive dyskinesia. There is also a close association between low levels of essential fatty acids and psychosis. Giving supplements of essential fatty acids to these patients produced a move towards normal in their red cell membranes, but they did not reach normal levels.

The overall results of this trial on tardive dyskinesia showed a small but significant improvement with *Efamol*. However, this study on human psychiatric patients was not as convincing as previous animal experiments which had shown that essential fatty acids do have a noticeable effect on abnormal movements.

Perhaps the reason why this trial had slightly disappointing results was because there are irreversible structural changes to the brain in patients with severe and prolonged movement disorders; perhaps the dose was too low and the treatment period too short. There may also be a problem with absorbing essential fatty acids; it probably takes much longer to correct abnormal levels of essential fatty acids in the brain, than in red blood cells. Further research is planned to see whether the results could be more positive with a higher dose over a longer treatment period.

Even though the results of this trial were only marginally significant for the role of evening primrose oil in tardive dyskinesia, it is still worth while giving supplements of evening primrose oil, together with neuroleptic drugs, in the prevention of management of this disease.

Memory and psychiatric status. The effect of *Efamol* as an anti-psychotic and as an improver of memory in tests was much more pronounced than its effects on tardive dyskinesia.

There was a mean improvement of 20% to 30% in the psychopathology scores at the end of the treatment with evening primrose oil. This improvement occurred in chronic patients with schizophrenia for whom orthodox neuroleptic therapy had little to offer. This related especially to the 'negative' symptoms of schizophrenia, such as 'flat' emotions and social withdrawal.

With memory, there was a clear deterioration when patients switched from active treatment to placebo. Other studies have shown that *Efamol* improves cerebral function to a significantly greater degree than a placebo.[8]

Vitamin E

It has been suggested that the chronic use of neuroleptic drugs may generate free radicals which damage synaptic terminals and destroy essential fatty acids. Vitamin E, which can restrict the formation of free radicals, produced an improvement in tardive dyskinesia.

It may be that the best results in schizophrenia, with or without tardive dyskinesia, may be achieved by combining essential fatty acids with Vitamin E. In the above study Vitamin E was added to both the *Efamol* and the placebo capsules.

Other nutritional approaches to schizophrenia: diet, zinc and B6

Cutting out wheat, milk and foods containing arachidonic acid (meat, dairy products, peanuts) has also been of help in some patients. According to Dr Carl Pfeiffer, most schizophrenics respond well to therapy with zinc and B6, in which they are often deficient. It could be that a combination of all these regimes is the most effective treatment.

Drugs are effective, particularly on the 'positive' aspects of schizophrenia, such as hallucinations and bizarre behaviour, but they do little for the 'negative' symptoms, the withdrawal and the lack of emotional contact. The nutritional approaches, in contrast, seem to do little for the 'positive' symptoms but help the 'negative' ones. There is therefore a case for the two approaches together.

CAUTION: In contrast to schizophrenia, temporal lobe epilepsy may be made much worse by supplementation with evening primrose oil.[9] Temporal lobe epilepsy and schizophrenia may be very difficult to distinguish from each other. However, the worsening of temporal lobe epilepsy in response to taking supplements of fatty acids means that giving EFA supplements could be the basis of a new diagnostic test. It is possible that many patients diagnosed as having schizophrenia in fact have temporal lobe epilepsy. If such patients are put on the appropriate drug, carbamazepine, they show a dramatic improvement which may mean they can leave hospital.

15 RAYNAUD'S SYNDROME AND SCLERODERMA

These two conditions are related in that patients with sclero-derma (also known as systemic sclerosis) often suffer from Raynaud's syndrome (or Raynaud's phenomenon).

Raynaud's syndrome

In general, Raynaud's syndrome affects the arteries in the fingers, and sometimes in the toes as well. No one knows exactly what causes it. The arteries contract in spasm, and as a result the fingers go pale and numb. Then they turn blue, and red in turn. They feel painful, tingling, and burning.

In severe Raynaud's syndrome (also commonly called Raynaud's phenomenon), the hands and feet turn abnormally cold and there may also be symptoms of a disease affecting the circulation such as atherosclerosis, or the connective tissue such as scleroderma (systemic sclerosis) or lupus erythematosus.

This condition can have serious side-effects. These include inflammation of the arteries of the fingers and toes, which could lead to a blood clot. A lack of blood in the extremities can lead to gangrene, or ulceration of the finger tips.

Scleroderma

Scleroderma is a rare disease affecting the body's connective tissue — the stuff that supports, separates and protects the organs of the body. The connective tissue progressively hardens and contracts. No one knows the cause, but it is thought to be an auto-immune disease, where the body attacks itself.

Strangely, it is three or four times more usual in women than in men. It most commonly strikes people in their thirties.

Scleroderma can affect any part of the body, including the skin, heart, kidney, lung or oesophagus. It can be localized, or it can spread slowly throughout the body.

Raynaud's syndrome is sometimes the first sign of scleroderma, and the patient may complain of abnormally cold hands and perhaps feet. Later on, other symptoms appear. In time, the fingers begin to swell, look like sausages, and can be difficult to move. As the disease progresses, the skin becomes thickened and taut, and movement becomes difficult. Scleroderma is a very serious condition, which can be fatal.

At the moment, both Raynaud's syndrome and scleroderma are incurable, and without any ideal treatment.

Evening primrose oil and Raynaud's syndrome

A study in Glasgow in 1985 showed that patients with Raynaud's syndrome definitely appeared to benefit from treatment with evening primrose oil.

The study, which was carried out at the Centre for Rheumatic Diseases at Glasgow Royal Infirmary,[1] found that six out of 11 patients felt a definite benefit from evening primrose oil, while two patients felt a moderate benefit, and three no benefit.

The group of 11 patients on the evening primrose oil treatment of 12 capsules a day were being compared against 10 patients on a placebo. As the weather got worse, the group taking the placebo experienced significantly more attacks than the group taking evening primrose oil. The severity of attacks and the coldness of hands improved in the evening primrose oil group.

One interesting result from this study was that if any improvement was going to happen at all, it was noticeable after four weeks of treatment. Sadly, there was a definite fall-off in the benefits by the end of the eighth week. Some improvements that were noted at two weeks or six weeks were not sustained beyond eight weeks. The peak results happened midway through the study.

However, all the results showed an improvement in the group taking evening primrose oil.

The study continued for only two months, and as yet there have been no studies using evening primrose oil for longer in these conditions.

The benefits of evening primrose oil

Many of the patients taking evening primrose oil said the course of treatment was like a tonic. Ten out of 11 patients on the evening primrose oil said they felt generally less depressed and had more energy. Seven patients asked to restart the drug two months later, complaining of a relapse in their Raynaud's symptoms and loss of mood elevation.

The doctors who conducted the study said 'in a disease such as Raynaud's phenomenon where the attacks can be precipitated by strong negative emotion, a drug with an anti-depressant effect might be useful.'

How evening primrose oil works in Raynaud's syndrome and scleroderma

Evening primrose oil was tried in Raynaud's syndrome because it makes PGE1 in the body naturally. PGE1 as a drug has been used effectively in Raynaud's phenomenon, showing an increase in hand temperature for up to six weeks.[2,3] In other words, the hands felt hotter. Normally, a patient suffering from Raynaud's syndrome has abnormally cold hands.

However, the problem with the drugs PGE1 and prostacyclin is that they have to be given intravenously and repeatedly in hospital. The hope was that evening primrose oil could achieve the same effect without the inconvenience. Indeed, the results of the Glasgow trial did show that evening primrose oil produced the same results as the infused drugs.

Apart from being a precursor of Prostaglandin E1, evening primrose oil has also been shown to increase the production of prostacyclin and to decrease levels of thromboxane B2, which constricts blood vessels. Also, evening primrose oil lowers vascular reactivity,[4,5] and may suppress chronic inflammation in animals.[6]

As an essential fatty acid, evening primrose oil also works in the body by helping to keep cell membranes fluid and flexible. This is an important function in a disease such as Raynaud's syndrome. Also, unsaturated fatty acids do not solidify in the cold — which is also an asset in Raynaud's syndrome. Patients who have both Raynaud's syndrome and scleroderma have found to be abnormally low in Prostaglandin E1.

Evening primrose oil and scleroderma [7]

The rationale for using evening primrose oil for scleroderma is the same as for Raynaud's syndrome.

The vascular complications of scleroderma may be successfully treated by infusions of prostaglandins which dilate the blood vessels. However, this treatment is impractical for long-term use.

A placebo-controlled study was carried out in which evening primrose oil or a placebo was given to patients. The researchers were looking for the fatty acid concentrations in plasma and red cells, and some particular prostaglandin concentrations in plasma.

Evening primrose oil produced a small but not significant rise in the concentrations of PGE1 and PGE2 and a significant fall in the concentration of thromboxane B2. Treatment also elevated the concentrations of dihomo-gammalinolenic acid and arachidonic acid in plasma. These changes went with clinical improvement.

In patients with scleroderma there seems to be some abnormality in the conversion of essential fatty acids to prostaglandins. Evening primrose oil may be able to help correct this.

16 OBESITY

Evening primrose oil does have an effect on obesity, but only when this is caused by metabolic abnormality. Unfortunately, it does not work in other types of obesity.

The discovery that evening primrose oil can help some obese people lose weight was made quite by chance. During a trial on evening primrose oil for schizophrenia at Bootham Park Hospital in York it was discovered that several patients who were more than 10% above their ideal body weight lost weight while taking evening primrose oil. There had been no changes to their diet.[1] The evening primrose oil had no effect on people who were within 10% of their ideal body weight. As a result of this chance finding, evening primrose oil began to be investigated as a treatment for metabolic causes of obesity.[6]

Brown fat (adipose tissue)

This is one of the key factors which explains why some people lose weight and others don't. The body has a special tissue known as brown fat which is found mainly in the back of the neck and along the backbone. The brown colour is due to the high concentration of cellular energy-producing (fat-burning) units called mitochondria. The brown fat burns calories not to produce energy for body movement, but solely for heat.

One role of brown fat is stabilization of weight; another is adaptation to cold weather. When brown fat is working normally it burns up any excess calories. But when brown fat is not working normally those calories are laid down as fat.[3] Some obese people have underactive brown fat, and this may be a metabolic disorder.

Interestingly, the essential fatty acid content of body fat is inversely proportional to body weight.[4] In other words, the higher the level of essential fatty acids in the body, the lower the body weight, and vice versa. A major study of over 600 men in Heidelberg, West Germany, found a strong inverse correlation between obesity, hypertension and serum cholesterol on the one hand, and the level of linoleic acid in adipose tissue on the other.[5]

The gammalinolenic acid in evening primrose oil had a stimulating effect on brown fat tissue. Also, the prostaglandins which are the end-products of evening primrose oil metabolism possibly accelerate the mitochondrial activity in the brown fat.

Metabolic causes of obesity and evening primrose oil

The Metabolic Unit at the University of Wales in Cardiff has been investigating evening primrose oil (*Efamol*) in severely obese patients.

The Cardiff doctors found that 30% of their patients had low sodium-potassium-ATPase activity, and they felt that this group had a metabolic cause for their obesity.[6] This complicated-sounding substance is an enzyme which is essential to the storage and transfer of energy in living cells and is found in all cell membranes. It is responsible for more than 20% of the total energy used by the body. Studies have shown that there are decreased concentrations of this enzyme in various tissues of obese mice. Together with defective brown fat, defective activity of the sodium-potassium-ATPase enzyme is a major cause of obesity.[6] Evening primrose oil activates this enzyme in cases where its activity is low. That's one reason why evening primrose oil was used in the study at the Metabolic Unit in Cardiff.

The effects of evening primrose oil were investigated in ten obese patients on a normal diet, against eight obese patients on a calorie-restricted diet. After six weeks, the evening primrose oil group lost an average of around 5 kilos, and their sodium-potassium-ATPase activity increased by an average of 52%. The group on the calorie-restricted diet lost an average of around 6½ kilos. Their ATPase fell by 37%.

The doctors concluded that through some unknown mechanism evening primrose oil might activate the mechanism involved in sodium-potassium-ATPase, so helping the body burn up excess calories.[2] The most effective dose used was eight capsules a day.

It must be emphasized that evening primrose oil is not helpful in the great majority of people who are overweight. It is useful only in the severely obese, usually only in those with a familial basis for their condition.

17 CANCER

One of the exciting possibilities for evening primrose oil lies in its use in treating cancer. The aim of using evening primrose oil in the treatment of cancer is to normalize tumours by stopping the cancer cells from proliferating, without affecting healthy cells.

This is very different from the existing orthodox approach that uses chemotherapy and radiotherapy, in which the treatment is toxic to both cancerous and healthy cells, with unpleasant side-effects. Treatment with evening primrose oil has no unpleasant side-effects.

Studies done during the 1980s all confirm that the GLA in evening primrose oil stops cancer cells proliferating, and can normalize them. Most of these studies have been in the laboratory, on animal and human cancer cells. However, there have been a few clinical studies on humans with various types of cancer.

Clinical studies on evening primrose oil and cancer

Preliminary studies in a few patients have shown promising results. One man with advanced lung cancer who was expected to live only four weeks survived for three years. Another man with bladder cancer, expected to live less than three months, was still alive and well three years later.

A doctor in South Africa [1,2,3] has reported excellent results in four patients with astrocytoma (a kind of blood cancer), and two with mesathelioma.

But perhaps the most interesting results so far have been from a pilot study on six South African patients with liver cancer.[4] In four patients, liver size decreased, and in two of these four patients this decrease was substantial.

Although five patients died, the mean survival time after diagnosis increased from 42 to 67 days. (This figure includes the survival time of two patients who were terminal on admisssion and died within days after the diagnoses were made.)

In all six cases symptomatic improvement was obvious, and quite striking improvement followed the start of the supplementation in three patients.

Only one patient entered the trial before the stage of severe clinical deterioration. He received *Efamol* for 172 days, and his clinical condition improved dramatically. This patient, a man aged 28, was admitted having been treated unsuccessfully for some time by a traditional healer.

After diagnosis of primary liver cancer, his diet was supplemented with six capsules a day of *Efamol* for the first 30 days. After that it was increased to 18 capsules a day, and 8g of Vitamin C a day were added as well. On day 127 *Efamol* was increased to 27 capsules a day, and the Vitamin C kept at 8g a day. He received no other treatment.

This patient was discharged from hospital, given a supply of *Efamol* and Vitamin C, and asked to return six weeks later. Tests showed that the tumour in his liver had gone down considerably.

Apart from any anti-tumour effects, most patients taking evening primrose oil for cancer have reported a substantially increased sense of well-being. The patients volunteered this information spontaneously; it was an unexpected bonus of the evening primrose oil treatment. For patients with cancer, an increase in well-being is a great asset.

Laboratory studies on evening primrose oil and cancer cells

Six different laboratories in four different countries have now obtained similar results: that polyunsaturated fatty acids normalize human cancer cells.[5,6,7,8,9,10,11] Tests have been done on at least nine different human malignant cell lines, including cancers of the liver, bone, oesophagus, breast, prostate, and skin. In all these tests, the normal cells remained unaffected.

Studies on animal cell lines using evening primrose oil have also had very good results. A study of breast cancer in rats showed that tumour growth was inhibited in rats given evening primrose oil.[12] These results agree with earlier findings [13] which showed that the growth of a mammary tumour was significantly reduced in rats treated with evening primrose oil.

The amount of oil given is important. The rats were fed a normal rat chow diet [12] containing 5% of total calories as fat. This was supplemented with 200 μl of evening primrose oil a day. The results showed that there was a significant inhibition of tumour growth up to 200μl, but greater amounts of oil began to increase tumour growth.

Other studies have shown that when rats are fed a diet containing 20% saturated fat, tumour growth increased. This may be because essential fatty acids cannot compete with so much saturated fat, with the result that they do not get metabolized properly.

There does seem to be a causal relationship between fat intake and the occurrence of breast cancer. However, it seems that the kind of fat as well as how much fat are important influences on the incidence of breast cancer. These studies on rats may have important lessons for human breast cancer.

Research in South Africa[6,7,8] (originally published in the *South African Medical Journal*) showed that gammalinolenic acid, taken from evening primrose oil, reduced cancer cell growth by up to 70%. GLA was added to three different types of malignant cell, both human and mouse. The mouse cancer cells were inhibited, and the human cancer cells taken from the oesophagus were killed. This research showed that although GLA was toxic to malignant cells, it had no such effect on normal cells.

The *Newsletter* of the Northwest Academy of Preventive Medicine[14] published these comments along with the findings: 'These data may have profound implications for the prevention and treatment of cancer. Whereas the usual method of fighting cancer is to destroy malignant cells, GLA may be capable of actually reversing, or retarding, the malignant process. Of prime importance is the fact that GLA is a normal metabolite and is essentially non-toxic.'

How evening primrose oil may be working in cancer

The active ingredient in evening primrose oil is gammalinolenic acid (GLA), and it is the GLA which gives the oil its cancer-controlling properties. It may be working in three different ways:

1 Lipid peroxides. What the researchers have found is that when human cancer cells are exposed to polyunsaturated fatty acids in the laboratory, the cells generate large amounts of substances called lipid peroxides, and die.[15]

Several different PUFAs have been tried to see which one is the most effective. GLA seems to be the best PUFA — it is highly toxic to malignant cells, but has no toxic effects whatsoever on normal cells.

It is now known that the majority of human and animal cancers contain low levels of lipid peroxides. There is evidence that lipid peroxides play a role in regulating cell division. In most cancers, the reason for the low levels of lipid peroxides seems to be a low concentration of PUFA substrate, which is needed for the formation of peroxide.

This may on the face of it seem paradoxical, because peroxides have a reputation for always being harmful. However, it now seems as though lipid peroxides, although indeed harmful in some situations, may be physiological regulators of cell division.

Some capsules of evening primrose oil contain Vitamin E as an anti-oxidant. However, studies have shown that Vitamin E in fact inhibits the toxic effects of GLA on malignant cells, so the best treatment for cancer would be large doses of evening primrose oil *without* the Vitamin E. Ideally, fish oils should also be taken. New clinical trials on cancer are using *Efamol Marine*, instead of *Efamol* on its own.

2 It by-passes the delta-6-desaturase enzyme block. Cancer is a known blocking agent of the metabolic pathway of linoleic acid. This block occurs at the first step, between linoleic acid and gammalinolenic acid, by inhibiting the delta-6-desaturase enzyme. The GLA in evening primrose oil by-passes this block by starting at the second stage in the metabolic pathway. This means that GLA can convert to DGLA and then to Prostaglandin E1 without hindrance.

3 Prostaglandins. Another way in which polyunsaturated fatty acids might be controlling cancer cells is by being converted into prostaglandins. Prostaglandins derived from PUFAs may inhibit the proliferation of human and animal tumour cells, and reverse transformed cells.

Originally it was thought that it was only PGE1 which had anti-cancer properties, but it now seems that other prostaglandins are important too.

Evening primrose oil as a potential treatment for cancer

It must be stressed that all the above studies are very preliminary, and although major clinical trials are planned, there are no results to date.

However, evening primrose oil is already widely used by alternative health centres and practitioners, such as the British Cancer Help Centre, for patients with cancer. In these cases it is used as part of a whole treatment programme, often involving several nutrients and a fundamental change of diet.

It must be emphasized that people with cancer should not treat themselves, but should always seek the advice of a qualified practitioner.

The studies with GLA are exciting, but the situation at the moment is that no double-blind trial has yet shown it to be effective. Until such trials are done, anyone with cancer taking evening primrose oil can only hope it will help them.

Note on AIDS: Some alternative practitioners use evening primrose oil as part of an intensive nutritional programme for AIDS sufferers. For further details, write to:
Nutrition Consultant,
Lamberts,
1 Lambert Road,
Tunbridge Wells,
Kent TN2 3EQ,
England.

18 COSMETIC USES — SKIN, EYES, HAIR, NAILS, AND BUST

Evening primrose oil is good for your skin. It helps make the skin smooth, counteracts dryness, protects against water loss, and may help slow down the ageing process.

The very first experiments on rats who were deprived of essential fatty acids showed what could happen to the skin and hair.[1] Sebum (grease) production increased, sebaceous gland size increased, and the skin broke out in red, itchy patches. The hair began to fall out, and the scalp to flake.

This knowledge led to the development of evening primrose oil as a treatment for eczema. But new research now shows that evening primrose oil can also help dry skin sufferers. There have been high hopes that evening primrose oil might help both acne and psoriasis, but this has not been proven as yet.

A dermatological test in Cardiff established that, with a dosage of eight to 12 capsules a day, there was a measurable improvement in skin roughness.[2] This result has recently been confirmed in Germany where doctors have again found that taking *Efamol* by mouth can actually make the skin smoother. This is the first time that any substance has actually been shown to do this.

How evening primrose oil works on the skin

The two major essential fatty acids, linoleic acid and gamma-linolenic acid, in which evening primrose oil is rich, are natural skin nutrients. Linoleic acid and gammalinolenic acid are vital components of the structure of all cell membranes and are normally converted by the body into prostaglandins. Prostaglandins play an important role in maintaining skin health.

When applied directly to the skin, linoleic acid and gammalinolenic acid have a profound effect on reducing transepidermal water loss, so evening primrose oil works as a natural moisturizer.

Unlike some other major organs of the body, skin is not capable of converting linoleic acid applied directly to it to gammalinolenic acid. The conversion of dietary linoleic acid to GLA and Prostaglandin E1 is known to decrease with age. As a consequence, the levels of GLA and PGE1 in the skin may become low, and the condition of the skin may suffer.

Because evening primrose oil is such a rich source of GLA, the problematic first step in the conversion process, from linoleic acid to GLA and through to PGE1, is completely avoided.

If you provide the skin with a direct source of GLA, it means that the liver — which normally does the conversion from linoleic acid to gammalinolenic acid — has less work to do. By applying evening primrose oil to the skin, you are providing it with a direct precursor of Prostaglandin E1 on site.

Evening primrose oil in cosmetics

It is not surprising that the cosmetic industry has cottoned on to the potential of evening primrose oil for a whole range of cosmetics. This is because evening primrose oil is an ideal ingredient of skin creams — it nourishes and moisturizes the skin, counteracts dryness, and protects the skin against water loss. Drying of the skin means loss of elasticity which can lead to signs of premature aging, so evening primrose oil may also prove useful in keeping the skin young and fresh.

Efamol, the company most responsible for the research and development of evening primrose oil, have a whole range of skin products under the brand name *Efamolia*. This consists of a moisture cream, a skin lotion, and an enriched night cream, all of which contain evening primrose oil, plus Vitamin A and Vitamin E.

Vitamin A has been shown to enhance the conversion of GLA to Prostaglandin E1. It also has important functions in the differentiation of epithelial cells during skin growth. Vitamin E has a protective function, shielding the skin from disturbing environmental influences such as sunlight.

The widest range of evening primrose oil cosmetics is from Creighton's, who do a vegetable soap, cleanser, toner, two moisturizers, a night cream, and some lotions, all containing evening primrose oil plus Vitamin E. Blackmore's also do a skin care range which includes evening primrose oil lotion.

The Body Shop has included evening primrose oil in its new 'Mostly Men' range of skin products. The fact that the evening primrose plant invades fields and gardens with a 'macho tenacity' is justification to use it in male cosmetics — the face wash, gel-based protective moisturizer, and facial scrub.

Several other companies, including some very classy brand names, use evening primrose oil in their cosmetics. The products can be found in health food shops, chemists, and the beauty counters of department stores.

Brittle nails, dry eyes, and dry mouth

The discovery that evening primrose oil can cure brittle nails was made quite by chance. A medical trial on evening primrose oil was being conducted in Scotland on two conditions which make the eyes and mouth dry and painful, Sjogren's syndrome and sicca syndrome.[3]

When people with these conditions were treated with evening primrose oil, it was found that not only did their dry eyes and mouth get better, but their brittle nails dramatically improved at the same time. The patients volunteered this information unsolicited.

In fact, people who are taking evening primrose oil for other conditions sometimes remark how good their nails have become since starting on the oil.

Results of a study for dry eyes and brittle nails

A small study conducted by Dr A. Campbell, consultant physician at Hairmyres Hospital, East Kilbride, and Dr G. MacEwan, consultant opthalmologist, Gartnavel General Hospital, Glasgow, came up with the following results, which were good for both dry eyes and brittle nails.

All the patients were treated with:
Efamol 500mg, 2 capsules, 3 times a day.
Pyridoxine (Vitamin B6) 25–50mg a day.
Vitamin C, 2–3g a day.

Patient 1. This 55-year-old woman was first seen in December 1979 with a six-month history of dry gritty eyes and brittle nails which were prone to split when she manicured them. On the above regime her eye condition gradually improved and after one month she spontaneously reported that her nails had become perfectly normal.

Patient 2. This 72-year-old women started on treatment in November 1979 following an eight-month history of dry, gritty eyes and a six-month history of nails splitting and breaking off. Within one month her eyes improved and by February 1980 they were moist and had normal tear secretion. At the same time her nails also became normal.

Patient 3. This 53-year-old woman had a history going back several years of defective tear secretion, and when she was seen in August 1980 she stated that her nails had been brittle and splitting for about six months. After one month of the above treatment her eyes had improved and her nails had stopped splitting.

Patient 4. This 48-year-old woman was diagnosed as having Sjogren's syndrome and rheumatoid arthritis. By December 1979, although her joint symptoms were variable and not particularly severe, her dry eyes and dry mouth had become marked and very troublesome. She had very brittle nails which could not be manicured and her hands and nails were very sensitive to exposure to detergents.

Therapy with non-steroidal anti-inflammatory aspirin-like drugs was stopped and she was started on the above regime. Over the next two months both her tear and saliva production improved and her nails, for the first time in many years, became very hard and could be manicured normally. At the same time the sensitivity to detergents disappeared.

In July 1980 therapy was stopped with the patient's consent. Her mouth became dry within one week and within two weeks she noticed that the skin around her nails had become dry and that her nails were beginning to split. Tear production seemed less affected. Treatment started again in September 1980 and again there was a rapid improvement in both her nail quality and the dry mouth.

In the light of these and other case histories, Drs Campbell and MacEwan suggested that the nutritional approach to dry eyes and brittle nails, using evening primrose oil, may be of value.

Dr Campbell also thinks that poor nail quality is a new sign of EFA deficiency in humans.

Hair

Animals deprived of essential fatty acids suffered from a dandruff-like condition of their fur, and loss of their coat. This suggests that essential fatty acids are needed for healthy hair.

Bust size

An unforeseen side-effect of evening primrose oil in some women is increased bust size. The women who have noticed this phenomenon remark that they did not put weight on anywhere else in their body. However, it seems that the effect of increasing bust size probably takes years to come about; no woman who has gone up in bra size has been taking evening primrose oil for less than a few years. No one knows quite what it is about evening primrose oil which helps the bust increase in size.

USEFUL ADDRESSES

Health support groups

Action for Research into
  Multiple Sclerosis (ARMS)
4a Chapel Hill
Stansted Mountfitchet
Essex CM24 8AG
tel (0279) 81555

American Allergy Association
PO Box 7273
Mento Park
CA 94026
USA

Huxley Institute and American
  Schizophrenia Association
1114 First Avenue
New York
NY10021, USA

Hyperactive Children's
  Support Group (HACSG)
59, Meadowside
Angmering
Littlehampton
West Sussex BN16 4BW
tel (0903) 725182
Secretary: Sally Bunday

International Women's Health
  Advisory Service
PO Box 31000
Phoenix
AZ 85046, USA

National Eczema Society
Tavistock House North
Tavistock Square
London WC1H 9SR
tel (01) 388 4097

National Women's Network
224 7th Street, S.E.
Washington D.C.
20003
USA

The Pre-Menstrual Tension
  Advisory Service
PO Box 268
Hove
East Sussex BN3 1RW
tel (0273) 771366

The Schizophrenia Association of Great Britain
Tyr Twr
Llanfair Hall
Caernarfon LL55 1TT
Wales
tel (0248) 670379
Hon. Sec.: Mrs Gwynneth Hemmings

Women's Health Concern
17 Earl's Terrace
London W8 6LP
tel (01) 602 6669
Run by Joan Jenkins

Manufacturers of evening primrose oil mentioned in this book

Efamol
manufactured by:
Efamol Ltd
Efamol House
Woodbridge Meadows
Guildford
Surrey
tel (0483) 578060

distributed by:

Britannia Health Products Ltd	Mr Ken Murdoch
Forum House	Nature's Way Products Inc
47-51 Brighton Road	PO Box 2233
Redhill	Springville
Surrey RH1 6YS	Utah 84663
tel (0737) 773741	USA

For further information about research data, please write to:

Efamol Research Institute
Annapolis Valley Industrial Park
PO Box 818
Kentville
Nova Scotia
Canada B4N 4H8
tel (902) 678 5534
The research librarian, Christina Toplack, has an exhaustive
collection of references on evening primrose oil.

Naudicelle
Manufactured and distributed by:
Bio-Oil Research Ltd
The Hawthorns
64 Welsh Row
Nantwich
Cheshire CW5 5EU
tel (0270) 629323

Distributor of Naudicelle in Australia
Key Pharmaceuticals
PO Box 12
Concord West
NSW 2138
Australia

Distributor of Naudicelle in New Zealand
Key Pharmaceuticals
3 Prosford Street
Ponsonby
Auckland, New Zealand

There are many other brands of evening primrose oil available in health food shops. Please write directly to the companies for product information, capsule analysis and information on bulk discounts.

Evening Primrose Oil is now available on the NHS in Britain as Epogram, manufactured by:

Scotia Pharmaceuticals Ltd
Woodbridge Meadows
Guildford
Surrey GU1 1BA
tel (0483) 574949

REFERENCES

1 What is the evening primrose?

1. Unger W. Fats and oil in the seeds of Oenothera biennis. Apotheker-Zeitung 1917;32:351–2.
2. Heiduschka A Luft K. The fat and oil in seeds of Oenothera biennis and one new linolenic acid. Archiv der Pharmazie 1919;257:33–69.
3. Eibner A Widenmayer L Schild E. Zur Frage der anstrich-technischeen Bedeutung der Isomerie der hoheren unge-sattigten Fettsauren und Glyzeride. Chemische Umschaurung 1927;34:312–26.
4. Riley JP. The seed fat of Oenothera biennis. Journal of the Chemical Society 1949;2728–2731.
5. Williams John. Bio-Oil Research Ltd, Historical Background. 1977.
6. Hassam AG Sinclair AJ Crawford MA. The incorporation of orally fed radioactive gamma-linolenic acid and linoleic acid into the liver and brain lipids of suckling rats. Lipids 1975;7:417–20.
7. Hassam AG Rivers JP Crawford MA. Metabolism of gamma-linolenic acid in essential fatty acid-deficient rats. Journal of Nutrition 1977;107:519–24.
8. Millar JH Zilkha KJ Longman MJ et al. Double-blind trial of linoleate supplementation of the diet in multiple sclerosis. British Medical Journal 1973 Mar 31;1:765–8.
9. Abdulla GH Hamadah K. Effects of ADP on PGE formation in blood platelets from patients with depression, mania and schizophrenia. British Journal of Psychiatry 1975;127:591–5.

2 From wild flower to crop

1. Stubbe W Raven PH. A genetic contribution to the taxonomy of Oenothera sect. Oenothera. Plant Systematics and Evolution 1979;133:39–59.

3 Essential fatty acids

1. Mead JF Fulco AJ. The Unsaturated and Polyunsaturated Fatty Acids in Health and Disease. Springfield IL: Charles C. Thomas, 1976.
2. Holman RT. Essential fatty acid deficiency. In: Holman RT, ed. Progress in the Chemistry of Fats and Other Lipids. NY: Pergamon Press, 1966:275–348.
3. Horrobin DF. Prostaglandins: Physiology, Pharmacology and Clinical Significance. Montreal: Eden Press, 1978.
4. United Nations — World Health Organization, Food and Agriculture Organization. Dietary Fats and Oils in Human Nutrition. Report of an Expert Consultation. Rome: UN-FAO, 1977.
5. Sinclair HM. Essential fatty acids and the skin. British Medical Bulletin 1958;14:258–61.
6. Burr GO Burr MM. A new deficiency disease produced by the rigid exclusion of fat from the diet. Journal of Biological Chemistry 1929;82:345–67.
7. Hansen AE. Role of unsaturated dietary fat in infant nutrition. American Journal of Public Health 1957;47:1367–70.
8. Hansen AE Haggard ME Boelsche AN et al. Essential fatty acids in human nutrition. Journal of Nutrition 1958;60:565–76.
9. Fleming CR Smith CM Hodges RE. Essential fatty acid deficiency in adults receiving total parenteral nutrition. American Journal of Clinical Nutrition 1976;29:976–83.
10. Brenner RR. Metabolism of endogenous substrates by microsomes. Drug Metabolism Review 1977;6:155–212.
11. Peluffo RO Ayala S Brenner RR. Metabolism of fatty acids of the linoleic acid series in testicles of diabetic rats. American Journal of Physiology 1970;218:669–73.
12. Bailey JM. Lipid metabolism in cultured cells. In: Snyder F, ed. Lipid Metabolism in Mammals II. NY: Plenum Press, 1977:352.
13. Kummerow FA. Nutrition imbalance and angiotoxins as dietary risk factors in coronary heart disease. American

Journal of Clinical Nutrition 1979;32:58–83.

14. Holman RT Aaes-Jorgensen E. Effects of trans fatty acid isomers upon essential fatty acid deficiency in rats. Proceedings of the Society for Experimental Biology and Medicine 1956;93:175–9.

15. Privett OS Phillips F Shimasaki H et al. Studies of the effects of trans fatty acids in the diet on lipid metabolism in essential fatty acid-deficient rats. American Journal of Clinical Nutrition 1977;30:1009–17.

16. Kinsella JE Hwang DH Yu P et al. Prostaglandins and their precursors in tissues from rats fed on trans-trans-linoleate. Biochemical Journal 1979;184:701–4.

17. Cook HW. Incorporation, metabolism and positional distribution of trans-unsaturated fatty acids in developing and mature lungs. Biochimica et Biophysica Acta 1978;531:245–56.

18. Hsu CM Kummerow FA. Influence of elaidate and erucate on heart mitochondria. Lipids 1977;12:486–94.

19. Yu P Mai J Kinsella JE. The effects of dietary 9-trans, 12-trans-octadecadienoate on composition and fatty acids of rat lungs. Lipids 1980;15:975–9.

20. Enig MG. Fatty Acid Composition of Selected Food Items with Emphasis on Trans-Octadecaenoate and Trans-Octadecadienoate. Ph.D. Thesis, University of Maryland, 1981.

4 Prostaglandins

1. Horrobin DF. Prostaglandins: Physiology, Pharmacology and Clinical Significance. Montreal: Eden Press, 1978.

2. Horton EW. The Prostaglandins. Berlin: Springer-Verlag, 1972.

3. Huttner JJ Gewbu ET Pangamala RV et al. Fatty acids and their prostaglandin derivatives: inhibitors of proliferation in aorta smooth muscle cells. Science 1977;197:189–91.

4. Gemsa D Seitz M Kramer W et al. The effects of phagocytosis, dextran and cell damage on PGE1 sensitivity and PGE1 production of macrophages. Journal of Immunology 1978;120:1187–94.

5. Lagarde M Dechavanne M Rigaud M et al. Basal level of platelet prostaglandins: PGE1 is more elevated than PGE2. Prostaglandins 1979;17(5):685–705.

6.   Casey CE Meydani S Walravens PA et al. Prostaglandins
      in human duodenal secretions. Prostaglandins and Me-
      dicine 1980;4(6):449–52.
7.   Karim SM Sandler M Williams ED. Distribution of
      prostaglandins in human tissues. British Journal of Phar-
      macology 1967;31:340–4.
8.   Minkes M Stanford N Chi MM et al. Cyclic adenosine
      monophosphate inhibits the availability of arachidonate to
      prostaglandin synthetase in human platelet suspensions.
      Journal of Clinical Investigation 1977;59:449–54.
9.   Horrobin DF. The regulation of prostaglandin biosyn-
      thesis: negative feedback mechanisms and the selective
      control of formation of 1 and 2 series prostaglandins:
      relevance to inflammation and immunity. Medical Hypo-
      theses 1980 Jul;6(7):687–709.
10.   Lands WE Byrnes MJ. The influence of ambient per-
      oxides on the conversion of eicosapentaenoic acid to
      prostaglandins. Progress in Lipid Research
      1982;20:287–90.

5 Premenstrual syndrome

1.   Horrobin DF Manku MS. Premenstrual syndrome: a
      disorder of essential fatty acid (EFA) metabolism. Pres-
      ented at: 2nd International Symposium on Postpartum
      Menopausal Mood Disorders, Kiawah Island SC, 1987
      Sep 9–13.
2.   Horrobin DF. Cellular basis of prolactin action: involve-
      ment of cyclic nucleotides, polyamines, prostaglandins,
      steroids, thyroid hormones, Na/K ATPases and calcium:
      relevance to breast cancer and the menstrual cycle.
      Medical Hypotheses 1979 May;5(5):599–620.
3.   Nazzaro A Lombard D Horrobin DF. The PMT Solution
      — Premenstrual Tension: The Nutritional Approach.
      London: Adamantine Press, 1985.
4.   Stewart M. Beat PMT through Diet. London: Ebury
      Press, 1987.
5.   Kerr GD. The management of the premenstrual syn-
      drome. Current Medical Research and Opinion
      1977;4(Suppl 4):29–34.
6.   Shreeve CM. The Premenstrual Syndrome. The Curse
      That Can Be Cured. Wellingborough: Thorsons Publ-
      ishers, 1983.

7. Brush MG Massil H O'Brien PM. The role of essential fatty acids in the premenstrual syndrome (Abstract). Presented at: 2nd International Symposium on Postpartum Menopausal Mood Disorders, Kiawah Island SC, Abstracts 1987 Sep 9–13:Abs 10.

8. Brush MG. Efamol (evening primrose oil) in the treatment of the premenstrual syndrome. In: Horrobin DF, ed. Clinical Uses of Essential Fatty Acids. Montreal: Eden Press, 1982:155–62.

9. Horrobin DF. The role of essential fatty acids and prostaglandins in the premenstrual syndrome. Journal of Reproductive Medicine 1983;28(7):465–8.

10. Ylikorkala O, Puolakka J, Makarainen L, Viinikka L. Prostaglandins and Premenstrual Syndrome. Prog Lipid Res. Vol 25, pp 433–435, 1986.

6 Benign breast disease

1. Pye JK Mansel RE Hughes LE. Clinical experience of drug treatment for mastalgia. Lancet 1985 Aug 17:373–6.

2. Pashby NL Mansel RE Hughes LE et al. A clinical trial of evening primrose oil in mastalgia (Abstract). British Journal of Surgery 1981;68(11):1.

3. Goolamali SK Shuster S. A sebotrophic stimulus in benign and malignant breast disease. Lancet 1975;1:428–9.

4. Horrobin DF. Cellular basis of prolactin action: involvement of cyclic nucleotides, polyamines, prostaglandins, steroids, thyroid hormones, Na/K ATPases and calcium: relevance to breast cancer and the menstrual cycle. Medical Hypotheses 1979;5(5):599–620.

7 Eczema

1. Bordoni A Biagi P Masi M et al. Evening primrose oil (Efamol) in the treatment of children with atopic eczema. International Journal of Clinical Pharmacology Research 1988:in press. And Biagi P Bordoni A Masi et al. A long term study of the use of evening primrose oil (Efamol) in atopic children. Drugs Under Clinical and Experimental Research 1988:in press.

2. Strannegard IL Svennerholm L Strannegard O. Essential fatty acids in serum lecithin of children with atopic dermatitis and in umbilical cord serum of infants with high or low IgE levels. International Archives of Allergy and

Applied Immunology 1987;82:422–23.

3.   Saarinen UM Katosaarir M Backman A et al. Prolonged breast feeding as prophylaxis for atopic disease. Lancet 1979;2:163–8.

4.   Midwinter RE Moore WJ Soothill JF et al. Infant feeding and atopy. Lancet 1982;1:339.

5.   Wright S Burton JL. Oral evening primrose seed oil improves atopic eczema. Lancet 1982 Nov 20;2(8308): 1120–2.

6.   Stewart JC Morse PF Moss M et al. Treatment of severe and moderately severe atopic eczema with evening primrose oil: a multi-centre study. 1988:in press.

7.   Schalin-Karrila M Mattila L Jansen CT et al. Evening primrose oil in the treatment of atopic eczema: effect on clinical status, plasma phospholipid fatty acids and circulating blood prostaglandins. British Journal of Dermatology 1987;117:11–19.

8.   Brenner RR. Nutritional and hormonal factors influencing desaturation of essential fatty acids. Progress in Lipid Research 1981;20:41–7.

9.   Parish WE Champion RH. Atopic dermatitis. In: Rook A, ed. Recent Advances in Dermatology. Edinburgh: Churchill Livingston, 1973.

10.   Byron NA Timlin DM. Immune status in atopic eczema. British Journal of Dermatology 1979;100:491–8.

11.   Warner JO Norman AP Soothill JF. Cystic fibrosis heterozygosity in the pathogenesis of allergy. Lancet 1976;1:990–1.

12.   Lloyd DH. Investigation of effects of essential fatty acid supplementation in atopic dogs (Summary of Paper). Presented at: Pre-Congress Dermatology Study Group, British Society of Veterinary Dermatology, BSAVA, London UK, 1987 Apr 2.

13.   Lloyd DH. Essential fatty acids and skin disease (Abstract). Journal of Small Animal Practice 1988:in press.

14.   Lloyd DH. Essential fatty acid supplementation in the management of skin disease in the dog and cat (Abstract). Proceedings of the European Society of Veterinary Dermatology, Berne, 1987 Jun.

8 Hyperactive children

1.   Colquhoun I Bunday S. A lack of essential fatty acids as a

possible cause of hyperactivity in children. Medical Hypotheses 1981;7(5):673-9.

2. Mitchell EA Aman MG Turbott SH et al. Clinical characteristics and serum essential fatty acid levels in hyperactive children. Clinical Pediatrics 1987 Aug;26(8):406-11.

3. Aman MG Mitchell EA Turbott SH. The effects of essential fatty acid supplementation by Efamol in hyperactive children. Journal of Abnormal Child Psychology 1987;15(1):75-90.

4. Case histories and data from the Hyperactive Children's Support Group. Secretary: Sally Bunday, 71 Whyke Lane, Chichester, West Sussex PO19 2LD.

9 Rheumatoid arthritis

1. Belch JJ Ansell D Madhok R et al. The effects of altering dietary essential fatty acids on requirements for nonsteroidal anti-inflammatory drugs in patients with rheumatoid arthritis: a double-blind placebo-controlled study. Annals of the Rheumatic Diseases 1988;47:96-104.

2. Horrobin DF. The regulation of prostaglandin biosynthesis: negative feedback mechanisms and selective control of formation of 1 and 2 series prostaglandins: relevance to inflammation and immunity. Medical Hypotheses 1980; 6:687-709.

3. Zurier RB Quagliata F. Effect of prostaglandin E1 on adjuvant arthritis. Nature 1971;234:304-5.

4. Karim SM Sandler M Williams ED. Distribution of prostaglandins in human tissues. British Journal of Pharmacology 1967;31:340-4.

5. Bach MA Bach JF. Effects of prostaglandins on rosette forming lymphocytes. In: Robinson HJ, Vane JR, eds. Prostaglandin Synthetase Inhibitors. NY: Raven Press, 1974:241-48.

6. Horrobin DF Manku MS Oka M et al. The nutritional regulation of T lymphocyte function. Medical Hypotheses 1979 Sep;5(9):969-85.

7. Shimizu T Rodmark O Samuelsson B. Enzyme with dual lipoxygenase activities catalyzes leukotriene A4 synthesis from arachidonic acid. Proceedings of the National Academy of Science (USA) 1984;81:689-96.

8. Zurier RB Ballas M. Prostaglandin E suppression in

adjuvant arthritis. Arthritis and Rheumatism 1973; 16:251–6.

9.   Krakauer K Torrey SB Zurier RB. Prostaglandin E1 treatment of NZB/W mice. Clinical Immunology and Immunopathology 1978;11:256–62.

10. Kunkel SL Ogawa H Ward PA et al. Suppression of chronic inflammation by evening primrose oil. Progress in Lipid Research 1981;20:885–8.

10 Diabetes

1.   Jamal GA Carmichael H Wier AI. Gamma-linolenic acid in diabetic neuropathy. Lancet 1986 May 10:1098.

2.   Houtsmuller AL van Hal-Ferwerda J Zahn KJ et al. Favourable influences of linoleic acid on the progression of diabetic micro- and macro-angiopathy in adult onset diabetes mellitus. Progress in Lipid Research 1982; 20:377–86.

3.   Howard-Williams J Patel P Jelfs R et al. Polyunsaturated fatty acids and diabetic retinopathy. British Journal of Ophthalmology 1985;69:15–18.

4.   Horrobin DF. The regulation of prostaglandin biosynthesis by the manipulation of essential fatty acid metabolism. Reviews in Pure and Applied Pharmacological Sciences 1983;4(4):339–83.

5.   Mercuri O Peluffo RO Brenner RR. Depression of microsomal desaturation of linoleic to gamma-linolenic acid in the alloxan diabetic rat. Biochimica et Biophysica Acta 1966;116:407–11.

6.   Peluffo RO Ayala S Brenner RR. Metabolism of fatty acids of the linoleic acid series in testicles of diabetic rats. American Journal of Physiology 1970;218:669–73.

7.   Faas FH Carter WJ. Altered fatty acid desaturation and microsomal fatty acid composition in the streptozotocin diabetic rat. Lipids 1980;15:953–61.

8.   Clark DL Hamel FG Queener SF. Changes in renal phospholipid fatty acids in diabetes mellitus: correlation with changes in adenylate cyclase activity. Lipids 1983;18:696–705.

9.   Jones DB Carter RD Haitas B et al. Low phospholipid arachidonic acid values in diabetic platelets. British Medical Journal 1983;286:173–5.

10. Pottathil R Huang SW Chandrabose KA. Essential fatty

acids in diabetes and systemic lupus erythematosus patients. Biochemical and Biophysical Research Communications 1985;128:803-8.

11. Brenner RR. Nutritional and hormonal factors influencing desaturation of essential fatty acids. Progress in Lipid Research 1981;20:41-7.

12. Jamal GA Hansen S Weir AI et al. An improved automated method for measurement of thermal thresholds. 1. Normal subjects. Journal of Neurology Neurosurgery and Psychiatry 1985;48:354-60.

13. Jamal GA Hansen S Weir AI et al. An improved automated method for measurement of thermal thresholds. 2. Patients with peripheral neuropathy. Journal of Neurology Neurosurgery and Psychiatry 1985;48:361-6.

11 Heart disease, vascular disorders and high blood pressure

1. Keys A Aravanis C van Buchem FS et al. The diet and all causes of death rate in the seven countries study. Lancet 1981;2:58-61.

2. United Nations — World Health Organization, Food and Agriculture Organization. Dietary Fats and Oils in Human Nutrition. Report of an Expert Consultation. Rome: UN-FAO, 1977.

3. Lewis B. Dietary prevention of ischaemic heart disease — a policy for the 80s. British Medical Journal 1980; 2:177-80.

4. Kannell WB. Meaning of the downward trend in cardiovascular mortality. Journal of the American Medical Association 1982;247:877-80.

5. Editorial. Why the American decline in coronary heart disease? Lancet 1980;1:183-4.

6. Katan MB Beynen AC. Linoleic acid consumption and coronary heart disease in the USA and UK. Lancet 1981;2:371.

7. Oliver M. Fats and atheroma. British Medical Journal 1979;1:889-90.

8. Yaari S Goldbourt U Even-Zohar S et al. Association of serum high density lipoprotein and total cholesterol with total cardiovascular and cancer mortality in a 7-year prospective study of 10,000 men. Lancet 1981;1:1011-4.

9. Rose G Shipley MJ. Plasma lipids and mortality: a source of error. Lancet 1980;1:523-5.

10. Wilson WS Hulley SB Burrows MI et al. Serial lipid and lipoprotein responses to the American Heart Association fat-controlled diet. American Journal of Medicine 1971; 51:491.

11. Turpeinen O. Effect of cholesterol-lowering diet on mortality from coronary heart disease and other causes. Circulation 1979;59:1–7.

12. Hoffmann P Taube C Ponicke K et al. Influence of linoleic acid content of the diet on arterial pressure of salt-loaded rats. Acta Biologica et Medica Germanica 1978;37:863.

13. MacDonald MC Kline RL Mogenson GJ. Dietary linoleic acid and salt-induced hypertension. Canadian Journal of Physiology and Pharmacology 1981;59:872–5.

14. Hoffmann P Taube C Forster W. Augmented acute hypotensive effect of dihydralazine and clonidine after linoleic acid rich diet in normotensive conscious rats. Prostaglandins Leukotrienes and Medicine 1982;8(4): 335–41.

15. Rao RH Rao UB Srikantia SG. Effect of polyunsaturate-rich vegetable oils on blood pressure in essential hypertension. Clinical and Experimental Hypertension 1981; 3:27–38.

16. Comberg HU Heyden S Hames CG et al. Hypotensive effect of dietary prostaglandin precursors in hypertensive men. Prostaglandins 1978;15:193.

17. Jacono JM Judd JT Marshall NW et al. The role of dietary essential fatty acids and prostaglandins in reducing blood pressure. Progress in Lipid Research 1982;20: 349–64.

18. Judd JT Marshall MW Canary J. Effects of diets varying in fat and P/S ratio in blood pressure and blood lipids in adult men. Progress in Lipid Research 1982;20:571–4.

19. Darcet P Driss F Mendy F et al. Fatty acid metabolism of plasma total lipid and platelet aggregation in old men given diet enriched with gamma-linolenic acid. Annales de la Nutrition et de l'Alimentation 1980;34(2):277–90.

20. Weber PC. Effects of evening primrose oil on platelet and plasma lipids and on platelet aggregation in middle-aged men and women. 1988:in press.

21. Fisher JM Donegan D Leon H et al. Effects of prostaglandins and their precursors in some tests of hemostatic function. Progress in Lipid Research 1982;20:799–806.

22. Malinow MR. The reversibility of atheroma. Circulation 1981;64:1-3.

23. Oliver M. Serum cholesterol: the knave of hearts and the joker. Lancet 1981;2:1090-5.

24. Christie SB Conway N Pearson HE. Observations on the performance of standard exercise test by claudicants taking gamma-linolenic acid. Journal of Atherosclerosis 1968;8:83-90.

25. Olsson AG Thyresson N. Healing of ischaemic ulcers by intravenous prostaglandin E1 in a woman with thrombo-angitis obliterans. Acta Dermato-Venereologica 1978; 58:467-72.

26. Kyle V Parr G Salisbury R et al. Vasospastic disease, cold stress and prostaglandin E. British Medical Journal 1981;2:1549.

27. Bierenbaum ML Oudhof JH. Platelet hyperaggregability in acute coronary disease managed with PGE1 (Abstract). International Prostaglandin Conference, Washington, Abstracts, 1979 May 10.

28. Horrobin DF Huang YS. The role of linoleic acid and its metabolites in the lowering of plasma cholesterol and the prevention of cardiovascular disease. International Journal of Cardiology 1987;17:241-55.

29. Horrobin DF Manku MS. How do polyunsaturated fatty acids lower plasma cholesterol levels? Lipids 1983; 18(8):558-62.

30. Hegsted DM McGandy RB Myers ML et al. Quantitative effects of dietary fat on serum cholesterol in man. American Journal of Clinical Nutrition 1965;17:282-95.

31. Keyes A Anderson JT Grande F. Prediction of serum cholesterol responses of man to changes in fats in the diet. Lancet 1957;2:959-66.

32. Siegel RJ Shah PK Nathan M et al. Prostaglandin E1 infusion in unstable angina: effects on anginal frequency and cardiac function. American Heart Journal 1984; 108:863-8.

33. Clifford PC Martin MF Dieppe PA et al. Prostaglandin E1 infusion for small vessel arterial ischaemia. Journal of Cardiovascular Surgery 1963;24:503-8.

34. Horrobin DF. A new concept of lifestyle-related cardiovas-cular disease: the importance of interactions between cholesterol, essential fatty acids, prostaglandin E1 and

thromboxane A2. Medical Hypotheses 1980 Aug;6(8): 785–800.

35. Enig MG Munn RJ Keeney M. Dietary fat and cancer trends — a critique. Federation Proceedings 1978;37: 2215–20.

36. Beare-Rogers JL Gray LM Hollywood R. The linoleic acid and trans fatty acids of margarines. American Journal of Clinical Nutrition 1979;32:1805–9.

37. Ohlrogge JB Emken EA Gulley RM. Human tissue lipids: occurrence of fatty acid isomers from dietary hydroge-nated oils. Journal of Lipid Research 1981;22:955–60.

38. Willis AL. Dihomogamma-linolenic acid as the endo-genous protective agent for myocardial infarction. Lancet 1984 Sep 22:697.

12 Multiple sclerosis

1.   Millar JH Zilkha KJ Longman MJ et al. Double-blind trial of linoleate supplementation of the diet in multiple sclerosis. British Medical Journal 1973 Mar 31;1:765–8.

2.   Field EJ Joyce G. Effect of prolonged ingestion of gamma-linolenate by MS patients. European Neurology 1978; 17:67–76.

3.   Bates D Fawcett PR Shaw DA et al. Polyunsaturated fatty acids in treatment of acute remitting multiple sclerosis (Letter). British Medical Journal 1978;2(6149):1390–1. And Bates D Fawcett PR Shaw DA et al. Trial of polyunsaturated fatty acids in non-relapsing multiple sclerosis. British Medical Journal 1977 Oct;2:932–3.

4.   Dworkin RH Bates D Millar JH et al. Linoleic acid and multiple sclerosis: a reanalysis of three double-blind trials. Neurology 1984 Nov;34:1411–5.

5.   Crawford MA Budowski P Hassam AG. Dietary manage-ment in multiple sclerosis. Proceedings of the Nutrition Society 1979 Dec;38(3):373–89.

6.   Swank RL Dugan BB. The Multiple Sclerosis Diet Book: A Low-Fat Diet for the Treatment of M.S. Garden City, NY: Doubleday, 1987.

7.   Horrobin DF Manku MS Oka M et al. The nutritional regulation of T lymphocyte function. Medical Hypotheses 1979 Sep;5(9):969–85.

8.   Field EJ Shenton BK Joyce G. Specific laboratory test for

diagnosis of multiple sclerosis. British Medical Journal 1974;1(905):412–4.

9.   Seaman GV Swank RL Tamblyn CH et al. Simplified red cell electrophoretic mobility test for multiple sclerosis (Letter). Lancet 1979 May 26;1(8126):1138–9.

10. Dunbar LM Bailey JM. Enzyme deletions and essential fatty acid metabolism in cultured cells. Journal of Biological Chemistry 1975;250(3):1152–3.

11. Bio-Oil Research Ltd. A statistical evaluation of Naudicelle as a dietary essential fatty acid supplement in Multiple Sclerosis patients.

12. Sanders H Thompson RH Wright HP et al. Further studies on platelet adhesiveness and serum cholesteryl linoleate levels in multiple sclerosis. Journal of Neurology Neurosurgery and Psychiatry 1968 Aug;31(4):321–5.

13 Alcoholism

1.   Horrobin DF Manku MS. Possible role of prostaglandin E1 in the affective disorders and in alcoholism. British Medical Journal 1980 Jun 7;280(6228):1363–6.

2.   Rotrosen J Mandio D Sagarnick D et al. Ethanol and prostaglandin E1: biochemical and behavioural interactions. Life Sciences 1980;26:1867–76.

3.   Nervi AM Peluffo RO Brenner RR. Effect of ethanol administration. Lipids 1980;15:263–8.

4.   Glen I Skinner F Glen E et al. The role of essential fatty acids in alcohol dependence and tissue damage. Alcoholism Clinical and Experimental Research 1987;11 (1):37–41.

5.   Glen E MacDonnell L Glen I et al. Possible pharmacological approaches to the prevention and treatment of alcohol-related CNS impairment: results of a double-blind trial of essential fatty acids. In: Edwards G, Littleton J, eds. Pharmacological treatments for Alcoholism, London: Croom Helm, 1984:331–50.

6.   Glen AI Glen EM MacDonald FK et al. Essential fatty acids in the treatment of the alcohol dependence syndrome. In: Birch GG, Lindley MG, eds. Alcoholic Beverages. London: Elsevier, 1985:203–21.

7.   Horrobin DF. Essential fatty acids, prostaglandins and alcoholism: an overview. Alcoholism Clinical and Experimental Research 1987 Jan/Feb;11(1):2–9.

8. Varma PK Persaud TV. Protection against ethanol-induced embryonic damage by administering gamma-linolenic and linoleic acids. Prostaglandins Leukotrienes and Medicine 1982 Jan;8(6):641–5.

9. Manku MS Oka M Horrobin DF. Differential regulation of the formation of prostaglandins and related susbstances from arachidonic acid and from dihomogamma-linolenic acid. I. Effects of ethanol. Prostaglandins and Medicine 1979 Aug;3(2):119–28.

10. Poisson JP Lemarchal P Blond JP et al. Influence of alloxan diabetes on the conversion of linoleic and gamma-linolenic acids to arachidonic acid in the rat in vivo. Diabète et Metabolisme 1978;4:39–45.

11. Vanderhoek JY Bryant RW Bailey JM. Regulation of leukocyte and platelet lip oxygenases by hydroxy eicosanoids. Biochemical Pharmacology 1982;31(21):3463–8.

14 Schizophrenia

1. Vaddadi KS Courtney P Gilleard CJ et al. A double-blind trial of essential fatty acid supplementation on abnormal movements, psychiatric status and memory in patients with tardive dyskinesia. Presented at: British Association of Psychopharmacology, Cambridge, UK, 1987 Jun.

2. Vaddadi KS. Penicillin and essential fatty acid supplementation in schizophrenia. Prostaglandins and Medicine 1979;2:77–80.

3. Vaddadi KS. Essential fatty acids in the treatment of schizophrenia. Presented at: World Congress of Biological Psychiatry, Stockholm, 1981.

4. Nohria V Vaddadi KS. Tardive dyskinesias and essential fatty acids: an animal model study. In: Horrobin DF, ed. Clinical Uses of Essential Fatty Acids. Montreal: Eden Press, 1982:199–204.

5. Vaddadi KS. Essential fatty acids and neuroleptic drug-associated tardive dyskinesia: preliminary clinical observations. IRCS Journal of Medical Science 1984;12:678.

6. Costall B Kelly ME Naylor RJ. The antidyskinetic action of dihomogamma-linolenic acid in the rodent. British Journal of Pharmacology 1984;83:733–40.

7. Kaiya H. Prostaglandin E1 treatment of schizophrenia. Biological Psychiatry 1984;19:457–63.

8. Besson J Glen E Glen I et al. Essential fatty acids, mean cell volume and nuclear magnetic resonance of brains of ethanol-dependent human subjects. Alcohol and Alcoholism 1987; Suppl 1:577–81.

9. Vaddadi KS. The use of gamma-linolenic acid and linoleic acid to differentiate between temporal lobe epilepsy and schizophrenia. Prostaglandins Leukotrienes and Medicine 1981 Apr;6(4):375–9.

15 Raynaud's syndrome and scleroderma

1. Belch JJ Shaw B O'Dowd A et al. Evening primrose oil (Efamol) in the treatment of Raynaud's phenomenon: a double blind study. Thrombosis and Haemostasis 1985;54(2):490–4.

2. Martin MF Dowd PN Ring EF et al. Prostaglandin E1 infusions for vascular insufficiency in progressive systemic sclerosis. Annals of the Rheumatic Diseases 1980; 40:350–2.

3. Belch JJ Newman P Drury JR et al. Intermittent prostacyclin infusions in patients with Raynaud's Syndrome. Lancet 1983;1:313–5.

4. Scholkens BA Gehring D Schlotte V et al. Evening primrose oil, a dietary prostaglandin precursor, diminishes vascular reactivity to renin and angiotensin II in rats. Prostaglandins Leukotrienes and Medicine 1982 Mar;8(3):273–85.

5. Begent NA Born GV Shafi S et al. Increased bleeding time associated with decreased vascular contracility in rats fed polyunsaturated lipids. Journal of Physiology 1983;349:68P.

6. Horrobin DF. The regulation of prostaglandin biosynthesis: negative feedback mechanisms and the selective control of formation of 1 and 2 series prostaglandins: relevance to inflammation and immunity. Medical Hypotheses 1980 Jul;6(7):687–709.

7. Horrobin DF Morse N Jenkins K et al. Plasma essential fatty acids and prostaglandins in patients with scleroderma (Abstract). 2nd International Congress on Essential Fatty Acids, Prostaglandins and Leukotrienes, Abstracts, London, UK, 1985 Mar 24–7:Abs 71.

16 Obesity

1. Vaddadi KS Horrobin DF. Weight loss produced by evening primrose oil administration in normal and schizo-phrenic individuals. IRCS Journal of Medical Science 1979;7(2):52.
2. Avenall A Leeds AR. Sodium intake, inhibition of Na/K ATPase and obesity. Lancet 1981;1:836.
3. Trayburn P Fuller L. The development of obesity in genetically diabetic obese mice pair-fed with lean siblings. Diabetologia 1980;19:148–53.
4. James WP Trayburn P. Thermogenesis and obesity. British Medical Bulletin 1981;37:43–8.
5. Osler P Arab L Schellenburg et al. Blood pressure and adipose tissue linoleic acid. Research in Experimental Medicine (Berlin) 1979;175:287–91.
6. Mir MA Morgan K Evans PJ et al. The effects of evening primrose oil (Efamol) on erythrocyte sodium transport and obesity. In: Horrobin DF, ed. Clinical Uses of Essential Fatty Acids. Montreal: Eden Press 1982:53–62.

17 Cancer

1. van der Merwe CF. The reversibility of cancer (Letter). South African Medical Journal 1984 May;65(5):712.
2. van der Merwe CF. Gamma-linolenic acid: a possible new cure for malignant mesothelioma (Abstract). 2nd International Congress on Essential Fatty Acids, Prostaglandins and Leukotrienes, Abstracts, London, UK, 1985 Mar 24–27:Abs 160.
3. van der Merwe CF. The use of gamma-linolenic acid in patients with various types of untreatable malignancies. A new nontoxic tool in the treatment of cancer (Abstract). 2nd International Congress on Essential Fatty Acids, Prostaglandins and Leukotrienes, Abstracts, London, UK, 1985 Mar 24–27:Abs 161.
4. van der Merwe CF Booyens J Katzeff IE. Oral gamma-linolenic acid in 21 patients with untreatable malignancy. An ongoing pilot open clinical trial. British Journal of Clinical Practice 1987 Sep;41(9):907–15.
5. Begin ME Das UN Ells G et al. Selective killing of human cancer cells by polyunsaturated fatty acids. Prostaglandins Leukotrienes and Medicine 1985;19:177–86.
6. Booyens J Engelbrecht P le Roux S et al. Some effects of

the EFAs linoleic and alpha-linolenic acids and their metabolites gamma-linolenic acid, arachidonic acid, eicosapentaenoic acid, docosahexaenoic acid, and of PGA1 and PGE1 on the proliferation of human osteogenic sarcoma cells in culture. Prostaglandins Leukotrienes and Medicine 1984;15:15–33.

7. Dippenaar N Booyens J Fabbri D et al. The reversability of cancer: evidence that malignancy in melanoma cells is gamma-linolenic acid deficiency-dependent. South African Medical Journal 1982;62:505–9.

8. Leary WP Robinson KM Booyens J et al. Some effects of gamma-linolenic acid on cultured human oesophageal carcinoma cells. South African Medical Journal 1982; 62:681–5.

9. Dippenaar N Booyens J Fabbri D et al. The reversability of cancer: evidence that malignancy in human hepatoma cells is gamma-linolenic acid deficiency-dependent. South African Medical Journal 1982;62(19):683–5.

10. Booyens J Dippenaar N Fabbri D et al. Some effects of linoleic acid and gamma-linolenic acid on the proliferation of human hepatoma cells in culture. South African Medical Journal 1984;65(15):607–12.

11. Booyens J Dippenaar N Fabbri D et al. The effect of gamma-linolenic acid on the growth of human osteogenic sarcoma and oesophageal carcinoma cells in culture. South African Medical Journal 1984 Feb 18;65:240–2.

12. Karmali RA Marsh J Fuchs C et al. Effects of dietary enrichment with gamma-linolenic acid upon growth of the R323OAC mammary adenocarcinoma. Journal of Nutrition Growth and Cancer 1985;2(1):41–51.

13. Ghayur T Horrobin DF. Effects of essential fatty acids in the form of evening primrose oil on the growth of the rat R323OAC transplantable mammary tumour. IRCS Journal of Medical Science 1981;9(7):582.

14. Newsletter of the Northwest Academy of Preventive Medicine 1983 Jan;8(1).

15. Begin ME Ells G Das UN et al. Differential killing of human carcinoma cells supplemented with n–3 and n–6 polyunsaturated fatty acids. Journal of the National Cancer Institute 1986 Nov;77(5):1053–62.

18 Cosmetic uses — skin, eyes, hair, nails, and bust

1. Sinclair HM. Essential fatty acids and the skin. British Medical Bulletin 1958;14:258-61.

2. Marshall RJ Marshall RW Holt PJ. Effects of Efamol on skin smoothness 1988:in press.

3. Campbell AC MacEwen CG. Systematic treatment of Sjogren's syndrome and the sicca syndrome with Efamol (evening primrose oil), vitamin C and pyridoxine. In: Horrobin DF, ed. Clinical Uses of Essential Fatty Acids. Montreal:Eden Press, 1982:129-38. And Campbell AC. Treatment of brittle nails with evening primrose oil (Efamol). In: Horrobin DF, ed. Clinical Uses of Essential Fatty Acids. Montreal:Eden Press, 1982: 125-8.

INDEX